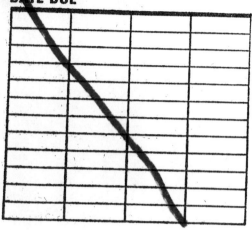

CAD:

Principles and Applications

CAD:

Principles and Applications

PAUL C. BARR

RONALD L. KRIMPER

MICHAEL R. LAZEAR

CHRISTOPHER STAMMEN

PRENTICE-HALL, INC., Englewood Cliffs, NJ 07632

Library of Congress Cataloging in Publication Data

Main entry under title:

CAD, principles and applications.

   Bibliography:
   Includes index.
   1. Engineering design—Data processing. 2. Computer-
aided design. I. Barr, Paul. II. Title: C.A.D.,
principles and applications.
TA174.C26 1985   620'.00425'02854   84-15937
ISBN 0-13-110198-6

Interior design: Shari Ingerman
Manufacturing buyer: Anthony Caruso

Printed in the United States of America

10 9 8 7 6 5 4 3 2 1

ISBN 0-13-110198-6 01

PRENTICE-HALL INTERNATIONAL, INC., *London*
PRENTICE-HALL OF AUSTRALIA PTY. LIMITED, *Sydney*
EDITORA PRENTICE-HALL DO BRASIL, LTDA., *Rio de Janeiro*
PRENTICE-HALL CANADA INC., *Toronto*
PRENTICE-HALL HISPANOAMERICANA, S.A., *Mexico*
PRENTICE-HALL OF INDIA PRIVATE LIMITED, *New Delhi*
PRENTICE-HALL OF JAPAN, INC., *Tokyo*
PRENTICE-HALL OF SOUTHEAST ASIA PTE. LTD., *Singapore*
WHITEHALL BOOKS LIMITED, *Wellington, New Zealand*

CONTENTS

PREFACE

CAD: Principles and Applications was motivated by the growing need to provide a basis and structure for the broad subject referred to as "CAD."

Computers have been used to facilitate the design process since the early 1960s. Applications have included analysis of designs, simulation, and even the complete automation of some of the design processes. In the early 1970s, interactive computer graphics became a practical tool for supporting graphics-based design. Presently, the great majority of computer-based workstations supporting design are of the interactive computer graphics type.

During the preceding twenty-year period the acronym CAD has been used to apply to any or all of the various application areas. *CAD: Principles and Applications* attempts to categorize CAD into its principal parts as we know it today from a computer graphics point of view: two-dimensional (2D) graphics or three-dimensional (3D) graphics; and from an applications point of view: general purpose or special purpose.

Following this theme, Chapter 1 presents an overview of the categories of CAD. The next 12 chapters present 2D and 3D general-purpose CAD functions that any system must perform. Variations on the terminology used in CAD systems to describe a feature or function and/or in the utilization or application of that feature or function are indicated as they emerge in the text.

One particular family of closely related CAD systems has been selected to serve as a unifying structure or thread for the development of our investigation of CAD. This system, which serves as a model, is a composite of several closely related systems. Although some of the features, functions, operations, or protocols might appear unique to this system, overall it is highly representative and fully illustrative of most other microcomputer-based 2D CAD systems.

Following the introductory chapter, Chapter 2 provides an extensive discussion of the applications of low-cost CAD systems in industry and as they are used in training.

Chapter 3 provides a general overview of the various common features of any 2D general-purpose CAD system. Specific common features, including general system characteristics and operations, as well as universal functions, are discussed briefly and related to the total operation of a 2D system.

Chapters 4 through 8 present information on the basic operational features and functions common to all 2D general-purpose drafting systems. By focusing on a composite model of several closely related systems, readers are guided through these primary features and functions and actually experience a sense of first establishing the parameters for a drawing using the UNITS and PROPERTIES Menus in the sample system; turning global features of a drawing, such as template lines, levels, and center markers, on and off using the SWITCHES Menu; creating primary objects using the ADD Menu; manipulating drawings in the workfile using FILER; establishing INPUT and OUTPUT parameters; using MODIFY to change primary objects; creating and then manipulating more complex drawings using the GROUP Menu; and using WINDOW to manipulate the views of a drawing by zooming in or out, and the like.

Chapter 9 presents three of the more advanced features of the sample 2D general-purpose drafting systems: LIBRARY, HATCH, and INQUIRE. The LIBRARY function enables the designer to create specific sets of symbols applicable to the special needs of a unique set of drawings (e.g., electrical or architectural symbols). The HATCH Menu in the sample 2D program provides a quick and easy method for automatically creating equally spaced parallel lines on any angle within any area bordered by an object or by text. The INQUIRE function allows the designer to analyze certain physical features of a drawing.

A 2D general-purpose drafting system can be productively employed in any number of disciplines including mechanical, electrical, and architectural drafting and design. Chapter 10 discusses a particular application of CAD in each of these disciplines.

Chapter 11 presents a general discussion of one introductory-level 3D general-purpose design system. The general-purpose system is a design tool for visualizing 3D objects; and 3D, when applicable, tends to automate both design and drafting.

Chapter 12 attacks the area of 3D special-purpose applications by addressing 3D piping design. It is apparent that a special-purpose application is of more limited scope but provides a greater opportunity for productivity gains since it relates specifically—in the design, selection, and utilization of features and functions—to only one area.

The final chapter, Chapter 13, addresses the development of a methodology for evaluating and selecting the various CAD systems presently available, or to be developed in the future.

The Appendix provides an extensive checklist of minimum features and functions that a good low-cost CAD system should have to be useful.

CAD: Principles and Applications can be used by industry to prepare for the task of selecting and installing a CAD system; or as a text for CAD training whether in industry or education. If used for training, the text is a good companion for the program operator's manuals and texts provided by any number of CAD computer systems vendors.

The authors of *CAD: Principles and Applications* wish to express their sincere appreciation to a number of people without whom the book could not have been written. Tom Lazear, President of T&W Systems, assisted with the editing and provided continued support and encouragement. Bob Murphy, also of T&W Systems, provided most of the original photos and artwork used to illustrate the text. John Staples, Rick Daber, and Eric Matson of Carrier Corporation also provided us with some photographs for the text and with other support. Bob Beauchemin, John Manifor, and J. Sunyogh of CADventures Unlimited permitted us to use many of their CAD-TUTOR photographs. Except where otherwise indicated, illustrations showing drawings in progress were drawn from CAD-TUTOR. Craig Tupper of CalComp provided technical assistance related to CAD systems and terminology. Ed Forrest, editor of *AEC Automation Newsletter*, provided valuable assistance in the development of our annotated bibliography. We would also like to thank the many instructors and students, too numerous to name individually, at colleges around the country for their input into the development of the book.

The authors owe a special debt to Ellen Denning of Prentice-Hall, who guided the manuscript to its successful completion.

Of course, any errors or omissions in the text are strictly the responsibility of the authors.

PAUL C. BARR
RONALD L. KRIMPER
MICHAEL R. LAZEAR
CHRISTOPHER STAMMEN

CAD:
Principles and Applications

1

INTRODUCTION

CAD, as used in this text, is the process of design when supported by computer methods. The objective of CAD is to increase productivity by utilizing computers in the design process. Although the term "productivity" has many meanings, it is used here to mean the ratio of labor hours required for a manual design function to the labor hours required if a computer is used to support the function.

CAD is a broad subject that fits into a broad spectrum of automated methods. The following diagram shows the place of CAD in the spectrum of automated methods:

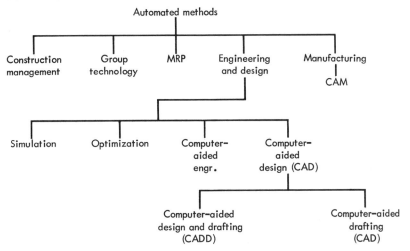

CAD systems may be either two-dimensional or three dimensional, and they may be designed to provide either special-purpose or general-purpose applications. Although there is no strict use of the acronyms, CADD generally refers to "3D" (three-dimensional) systems that build a model of a design and tend to automate drafting. Computer-aided drafting, or computer-aided design (CAD), generally refers to "2D" (two-dimensional) systems that support drafting more directly.

Think of 2D as a computer system that replaces the T-square, triangle, template, pencil, and eraser. 2D is to the draftsman what word processing is to the secretary. The designer constructs the design and then commits the design to paper through the drafting process.

A 3D system is one step ahead in the design process. A model of the design is put in the computer and the computer tends to automate the drafting process.

A "special-purpose" application refers to a system dedicated to a specific narrow application, such as PCB boards, piping, and the like. Special-purpose CAD systems are often referred to as vertical systems.

A "general-purpose" application is a more general application such as a higher-level language that can be applied in a number of application areas. General-purpose CAD systems are often referred to as horizontal systems.

CAD systems, then, can be categorized in this matrix:

C A D

	General	Special
2D		
3D		

The great majority of CAD systems in use today are general-purpose 2D systems that are being applied to help drafting in any discipline: architectural, mechanical, civil, electrical, or electronic.

All CAD systems that in any way aid or automate drafting have the following components:

- Computer
- Graphics screen
- Graphics input such as a digitizer
- Graphics output such as a plotter

Figure 1.1 is a good illustration of the components of a typical low-cost CAD system.

Attempts have been made to list and categorize CAD products. Table

Figure 1.1 Typical low-cost microcomputer-based CAD station. (Courtesy of APPLE Computers and T&W Systems.)

1.1 is offered as indicative of the sorts of things that CAD systems in particular are being used to produce. The list is ordered, more or less, from easy 2D applications that a low-cost system can handle to more complex applications which are more difficult to implement on a low-cost system.

TABLE 1.1 CLASSES OF CAD REQUIREMENTS

Type	2D or 3D	To scale?	Dimensioned drawing?	Examples
1	2D	No	No	Schematics, wiring diagrams, organization charts
2	2D	Yes	No	Technical illustration, layouts (floor equipment, ads)
3	2D	No	Yes	Design sketches
4	2D	Yes	Yes	Architectural and mechanical drafting
5	3D	No	No	3D schematics (e.g., material flow)
6	3D	Yes	No	Architectural graphics, technical illustration, parts explosion, automatic design
7	3D	No	Yes	Piping isometrics, conduit isometrics
8	3D	Yes	Yes	Part design, tool path design

CAD AND PRODUCTIVITY

One of the most important concepts related to CAD that needs definition is the concept of productivity. Although many benefits of CAD are listed by proponents—such as quality of drawing, shorter design cycle, employee and customer satisfaction, less paper, and the like—the bottom line for CAD is always productivity. Generally, productivity is thought of either implicitly as improvement in labor hours required to produce a product; or conversely, as the increase in product produced by a given staff. For example, the increase in gross national product (GNP) in the United States from a given population would be a measure of productivity increase.

The problem with looking at productivity solely on the basis of labor hours is that it does not take into account the cost of the tools required to produce the labor-hour improvements. So what we need to do is to define a benefit ratio of the cost to produce a design manually to the cost to produce the design with a computer. When that is done, the "all-in" cost of the workstation (possibly including some central computer costs) is a very important parameter. For example, if a particular CAD application yields a productivity ratio of 1.5, the workstation must cost less than $40,000 (allowing for approximate salary and reasonable fringe benefits for a designer) or the application will not pay out (i.e., have a benefit ratio greater than 1.0). So, in addition to saving labor hours, the CAD application must meet economic criteria or little, if anything, will have been saved.

CAD PRODUCTIVITY ILLUSTRATED

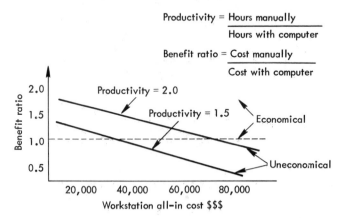

In considering the productivity and cost benefits to be derived from a low-cost CAD system, certain caveats are in order. An Apple-based (or many microcomputer-based) system will not do everything that a Computer-vision, CADAM, CalComp, or Intergraph system will do—although some of the very sophisticated microcomputer-based systems come very close. As a rule, larger systems are capable of handling more complex CAD software programs with greater speed and have some functional enhancements beyond their smaller kin. For example, a larger system can handle single drawings with up to 10,000 objects; some Apple-based systems handle no more than 2000 objects conveniently. (*Note*: The complexity of drawings is measured in "objects" or "entities." An object or entity may be a single primitive, a more complex grouping of individual primitives, a symbol, a string of text, a dimension, or the like.)

IMPLEMENTATION ALTERNATIVES

The literature on CAD presents a considerable number of implementation options. This large number, however, can be divided into more manageable units.

Essentially, there are four categories of options: using a CAD service bureau, adding software and terminals to an existing in-house maxicomputer, buying a turnkey minicomputer package from one of a multitude of good vendors, or experimenting with a personal (micro) computer and micro-based CAD software. Table 1.2 illustrates these options.

The service bureau makes a lot of sense for the first-time engineering user. The company can buy only the services needed. The service bureau develops the software and buys the hardware—a good way to learn. The disadvantages are high cost at high volumes and inflexibility with regard to modifications and tailoring of the software.

TABLE 1.2 REPRESENTATIVE ALTERNATIVES FOR IMPLEMENTATION CAD SYSTEMS

| Service bureau | Turnkey | | Personal |
	Maxi-based	Mini-based	
McAuto-Unigraphics	CDC	AM Bruning	Apple
Compaid	IBM	Applicon	IBM
CDC Cybernet		Autotrol	Radio Shack
Boeing		Bausch & Lomb	
Lockheed CADAM		Calma	
Ditechs		CalComp	
Master Design		Carrier	
		Computervision	
		Hewlett-Packard	
		Holguin	
		Intergraph	
		Prime	
		Tektronix	
		Terak	
		T&W Systems	

Adding software and hardware to an in-house computer always seems like a good idea. Certainly, there are advantages to consider: in-house support, use of excess CPU capacity, and a big capacity for data-base handling. Potential problem areas to watch for include: Will the capacity be there to support a real-time operation? Is there good CAD software available for the existing mainframe computer system? What is the terminal response time? What is the real, "all-in" workstation cost?

By far the most popular solution is the turnkey hardware/software package. Generally, these packages are implemented on minicomputers that are dedicated to the CAD function. To the degree that the available software fits a problem and to the degree that the problem is divisible to fit on several computers, this solution can be both effective and attractive.

The recent development of the microcomputer-based CAD system provides the fourth alternative. Some very creative things are happening here, and for certain applications, the microcomputer, with a monitor, digitizer, and CAD software can do a better-than-adequate job at a very low cost. But there currently are limitations to this solution. At present, most microcomputers for which CAD software has been developed, are "small" computers, with comparatively limited computational power and storage capacity, and they cannot yet handle large applications.

COMPARISONS

So how does one decide which of the categories to choose and which vendor or approach to select within that category? There are no magic answers. Systems are expensive both in direct cost and in follow-on costs such as training, and there are risks in choosing the wrong system for a given application. Also, new technology is being introduced at a very rapid rate. The only way to assure a correct decision is to carry out a thorough advisability study, as described in the next section. There are, however, some broad characteristics of each category, and these are outlined in Table 1.3.

THE ADVISABILITY STUDY

Purchase or development of a computer graphics system requires careful analysis and consideration. There are lots of options to choose from and there are some large risks—as well as substantial gains. Described here is a procedure following the SPECTRUM (4) approach that has been used in a number of instances. Following this, or some similar procedure, increases the probability of making the right choice. (A more extensive discussion of evaluating and selecting a CAD system appears in Chapter 13.)

TABLE 1.3 SYSTEMS COMPARISON

	Service bureau	In-house maxi	Dedicated mini	Micro-computer
Example Applications	3D CAD Structural Piping isometrics	3D CAD Structural Piping isometrics MIS graphic	IC design Structural Building layout Flowsheets Diagrams Tool design Piping isometrics	Small Calculations Pictorial
Data base	Good, but $$$	Integrated	Distributed	Limited
Workstation all-in cost per hour	$100–$150	$50–$100	$10–$40	$5–$15
3D system development cost	Training	$500–$20,000	$500–$20,000	$10–$1000
Software package availability	Good	Good	Excellent	Good
Response	Good	Depends	Good	Fair
Reliability	Good	Depends	Good	Good
Maintenance	By others	Good	Good	Good

The first step in the study should be to develop a clear definition of requirements. Study the existing system and develop a checklist of features that should be included in the new system. Then, use the checklist to interview the engineers, designers, and management who will be affected. Based on the interviews and subsequent analysis, the second step is to write a specification of how the system should operate in your environment. Consider such things as how the system will fit in your paper-flow scheme, how it will be organized, where it will be placed, screen response times, complexity, functions it will perform, data required, how it should interface with other systems, accuracy, quality of the drawing, type of input, and the like. Review the specifications with the people involved.

The third step is to choose a short list of alternatives that might solve the problem considering the four solution options or alternatives—service bureau, in-house mainframe, minicomputer, microcomputer—as well as system scope. Selection of a short list usually involves screening out alternatives that do not meet a given minimum economical advantage. Here you may use any financial selection criterion.

In this example, we would screen out any alternatives that have an estimated productivity gain and machine cost that yields a benefit less than 1.0.

Fourth, develop criteria for selection based on the specification and match that against each of the alternatives to be considered. Now, go back to the people interviewed and ask them to comment on the alternatives to be considered and the criteria for selection. Only after they have agreed to the basis for selection should an attempt be made to make a selection. This avoids the win–lose situation of the systems people trying to decide what is best and then trying to "sell" that to the engineers and designers who must use it.

The fifth step is to collect information about each alternative and to complete the analysis of each against such criterion. Then, make a recommendation and go back again to the people involved for confirmation.

It may be necessary to conduct this study in phases. In the first phase, you might decide on a possible vendor. For example, the first phase of the study might lead to the selection of the turnkey minicomputer approach over the service bureau or the in-house mainframe alternatives. Then it would be necessary to repeat step five by preparing a formal "request for proposal" (RFP) from the specifications developed in step two.

When complete, review the study results again with members of management. The selection of a computer-aided engineering system is a complex process. In addition to the technical and economic factors, which are similar to any other capital purchase, there are also human factors, which will affect the way people view their work. The way to make those decisions is to involve all personnel who will be directly affected by such a decision in the collection of data, and then to refer all data to a group of competent managers to review and analyze all the facts so that a consensus decision may be reached.

CONCLUSION

Computer graphics offers one possible solution to some of the problems faced by engineering companies today. Technological advances have improved systems so that today there are a large number of alternatives to choose from. On the one hand, this provides great opportunities for increased creativity, productivity, and profits. On the other, it increases the complexity of the decision-making process and of the risks involved. A careful review of all alternatives with a broad selection of employees and management is the recommended approach to use to ensure success.

The remaining chapters of this book are intended to impart a good understanding of how a modern CAD system works to help prepare a manager for the process of evaluating and installing a CAD system to meet unique needs, or for the actual operation of a CAD system. The book is organized around the categories of CAD systems presented in this chapter: 2D, 3D, general purpose, special purpose. The first that is presented is 2D general purpose since that category represents the majority of systems available today. General-purpose 3D and piping as an example of special-purpose 3D are discussed more briefly. Finally, a fairly thorough discussion of selecting and implementing a CAD system is provided in Chapter 13.

2

APPLICATIONS OF LOW-COST 2D CAD SYSTEMS

The basic problem facing industry in the coming decades is one of enhanced productivity. One means to improve productivity is through the introduction of further automation, in both the manufacturing and design areas. But to effect a significant improvement in productivity, a number of problems must be solved and needs satisfied, including overcoming the high cost of installing a CAD system, especially for smaller production environments, and the cost of training.

For example, the majority of architects work in shops of 10 or fewer persons, and a large number of engineers, freelance designers, and draftsmen work in similar shops. These companies cannot afford to buy or to support a CAD system with a six- or seven-digit price tag that requires special environmental controls and a full-time computer operator—even if the system does increase productivity by 2 to 1, or more.

There is also the cost of training—approximately $10,000 for an operator on a contemporary CAD system. There is, in addition, often a repetitive cost factor for training, since each new software release might require additional hours of training. This is a significant obstacle for any small company that might wish to look toward CAD for increased productivity and enhanced competitiveness in the marketplace.

APPLICATIONS OF LOW-COST CAD

Use of microcomputers for drafting and design is an application of the 80-20 rule; that is, while providing 80% or so of the functions of the higher-cost systems, the cost is only one-fourth or one-fifth as much as that of a major CAD system. While applications are increasing constantly, low-cost CAD is usefully employed in CAD training, as preprocessors for more complex and expensive systems, for creating less-demanding drawings, and for applications that simply do not pay out on the larger and more expensive CAD systems. In addition, the microcomputer in the low-cost CAD station is suitable as an engineering workstation for general-purpose applications ranging from word processing to scheduling, electronic spreadsheet analysis, structural analysis, and more.

CAD training is perhaps the low-cost CAD application of highest potential. Estimates indicate that 20,000 CAD workstations are being installed per year, and reductions in cost per station might well increase this estimated number. That means that there is a demand for training over 20,000 new CAD operators, technicians, design engineers, and analysts per year. In addition, the thousands of engineering students graduating each year need to be trained so that they are prepared to work in a CAD environment.

The cost to train an operator of a sophisticated 2D or 3D system has been estimated at $10,000, half of which is the cost of using the sophisticated system in the training—lost productivity and system downtime. That means that $1 billion or so is spent on CAD training every five years. Substitution of a low-cost CAD station in industry, or the installation of low-cost CAD stations in vocational and engineering schools with appropriately

Figure 2.1 Typical low-cost microcomputer-based CAD station. (Courtesy of APPLE Computers and T&W Systems.)

designed industry-based curriculum, pays the cost of each low-cost system after only two or three students. Furthermore, some experiments indicate that students learn more quickly on a small system because it is less intimidating and they can quickly "get their arms around" the system. New developments in microcomputer architecture and operation make this point of view increasingly appealing.

Microcomputer stations are already in use as CAD preprocessors. A system produced by Avera is a preprocessor for PCB applications. Systems produced by T&W Systems are used as preprocessors for larger DEC 10 systems and for Terak-based systems. The primary advantage of the preprocessor is that it permits offloading the more routine data-entry operations. Since use of the low-cost preprocessor can be learned more quickly, managers can be surer that new operators are right for the job; and new operators can be surer that they will like the job as training progresses. The result: reduced training costs, less risk of a mismatch of talent and job, and the opening of a career path for employees.

Some types of CAD products can be produced more easily than others. The easy CAD applications, such as overheads, schematics, and illustrations, are naturals for the lower-cost CAD systems. It makes no economic sense to produce a CAD product on a computer system that is any more expensive than is necessary to do the job.

At the same time, there are some more difficult CAD applications that yield a labor-hour-based productivity ratio of 1.5 or so; process and mechanical flow diagrams are just two such examples. To yield an economic benefit (i.e., a benefit ratio greater than 1.0) it is necessary that this class of applications be run on a workstation that costs less than $40,000, which is the cost of a designer, including reasonable and typical fringe benefits. This is a case where the system designer must design-to-cost or the system actually provides a disservice to the company.

The same microcomputers that are used as preprocessors in a low-cost CAD system are also available to the user for general-purpose computer applications. The excellent operating systems available on microcomputers today—CP/M, UCSD Pascal, UNIX, and others—have paved the way for some very useful software on micros. VisiCalc is the most successful software package ever introduced—and ever imitated—having sold over 100,000 copies. The Milestone critical path scheduling system is not far behind. VisiCalc will operate on most microcomputers without any enhancement to their memory or operating system; Milestone simply requires that UCSD Pascal be implemented. SAP80, one of the most popular finite-element structures systems, has been implemented by the University of California, Berkeley, on a number of microcomputers, including those using CP/M and the DEC microcomputers, which use the RT-11 operating system. Other software available, of course, includes general business, word processing, statistical and math, regression, and project management applications.

The microcomputer-based CAD system is being applied today in CAD training, preprocessing for larger systems, for less-demanding CAD applications, and for general-purpose computing. Although low-cost CAD systems cannot do everything that the larger and more sophisticated—and more costly—systems can do, they have a well-established foothold and their promise for the future is even greater. In the next few years the increasing power of personal computers will allow these systems to handle more complicated drafting tasks. Better computer power will be complemented by size and resolution improvements in graphic screens.

The problem in choosing a low-cost CAD system almost always boils down to applications software. Hardware by itself, although important, is useless without software. Incomplete or poorly written software restricts the overall usefulness of the CAD system. The following section offers some considerations of the functions and features that should be required of a CAD software package.

REQUIREMENTS OF LOW-COST CAD SOFTWARE

A low-cost general-purpose interactive drafting software package should be designed for microcomputers to efficiently increase the designer's labor productivity at a price that provides for economic productivity as well. The low-cost CAD system would be general-purpose and would have application in any design environment, including architectural; civil, mechanical and electromechanical drafting; instrumentation diagrams; flow diagrams; technical illustrations; ads; printed circuit boards; facilities and floor layouts; plate and sheet metal drafting; and graphic arts.

The low-cost CAD system should be able to create a symbolic drawing using a graphics device or keyboard for input, a graphics screen for review and editing of the drawing, and an X-Y plotter for placing the drawing on bond, vellum, or Mylar. The system should use a low-cost, medium-resolution rastar scan graphics CRT to create and manipulate a rough sketch of the final drawing. The drawing can then be drawn on a high-quality, high-resolution output device such as a plotter.

The low-cost system should be easy to learn and to use, and should be designed with a twofold goal in mind: to help the novice designer to proceed quickly up the learning curve in CAD, while providing the experienced designer with all the needed capabilities of a CAD system. CAD systems that are menu driven are the easiest to learn. All available options should be displayed and selected by a single keystroke. An example of such a menu-driven system is illustrated in Figure 2.2

Messages that are displayed should be clear and concise and should direct the designer by prompting responses. Many low-cost CAD systems provide a "Help" feature which the designer can use as a quick reference

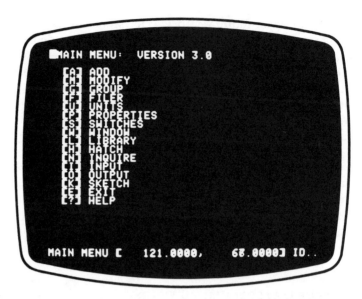

Figure 2.2 Typical 2D general purpose CAD system Main Menu with options displayed.

guide. The "Help" feature should display a condensed portion of the reference manual on the screen.

The low-cost CAD system should be capable of automatically dimensioning a distance on the drawing. Leader lines and arrowheads should be automatically created and then placed under the designer's control. The numeric distance calculated by the software should then be placed near the dimension leader lines by the designer.

A variety of input modes for coordinate data should be provided by the system, including one to allow the designer to graphically input coordinate data on the drawing and to allow coordinate data to be input from the keyboard. All coordinate data should be in "real-world coordinates" with the CAD system performing automatic scaling.

It should be possible to interactively modify any part of the drawing by moving, copying, imaging, rotating, scaling, and deleting individual objects. The object to be modified should be selected simply by touching it with the cursor.

The designer should also be able to define and manipulate groups of objects (e.g., predefined symbols) on the drawing. They, too, should be easily moved, copied, imaged, rotated, scaled, or deleted from the drawing. Groups of objects can be used to represent such symbols as a resistor, a pressure valve, or a door.

The various properties, attributes, or characteristics of both objects and groups should be under the complete control of the designer. Properties such as linestyle, pen color, density, group name, and level, should all be easily changed through a simple sequence of keystrokes.

At least 30 independent levels should be provided by the program. Any combination of these levels, consisting of objects on the drawing, should be available for display by the designer. For example, the floor plan, electrical conduiting, furniture, and notes of an architectural drawing can all be placed on separate levels.

When the designer is working on a drawing, that drawing should be only a copy of the original drawing. This copy should be temporarily stored as a workfile drawing. The original drawing is changed only when the working drawing is saved on diskette. The workfile concept should allow complete recovery of all work in case of power failure or other computer malfunction.

A complete set of file management options should be provided so that the designer can save, retrieve, and delete drawings to and from diskette. A directory option should provide a listing of drawings that are on a diskette. Also, a complete set of workfile maintenance options should be provided.

The designer should be able to "zoom in" to inspect and modify magnified sections, as well as to "pan" across the drawing surface. There should be no practical limit to the number of times that the designer can perform these windowing—zoom and pan—functions.

A complete grid system should be provided, allowing for independent X-axis and Y-axis grid spacing as well as grid subdivisions.

There should be various snap modes, such as increment and grid snap. The first, increment snap, automatically locks the cursor to the nearest increment (an increment is defined as the smallest measurement used in the drawing). The second, grid snap, automatically locks the cursor to the nearest grid intersection. Snap helps the designer to keep the drawing properly aligned.

The low-cost CAD system should be interfaced to most of the popular A, B, C, and D size plotters. Features should allow the designer to precisely position and plot the drawing—or any part of the drawing—to any scale.

Even a low-cost CAD system should be capable of maintaining graphics information (coordinate data) to at least seven digits of accuracy. For example, the drawing of a machine part of 100.0625 inches should be accurately stored. A low-cost CAD system should also have sufficient capacity to handle drawings containing as many as 4000 objects.

The advent of relatively low cost 2D CAD systems has made CAD available to more schools; and the recognition by industry that such systems have a valuable role to play in education—and even in industry—has led to the rapid growth in the number of systems installed in universities, colleges, community colleges, technical institutes, and other postsecondary schools, and in the development of additional (and more sophisticated) low-cost microcomputer-based systems. Low-cost 2D CAD can contribute toward the need for CAD training and to the initial entry of a company into CAD, and should not be overlooked.

The several chapters that immediately follow provide a detailed description of some of these basic functions, features, and operations of a typical low-cost microcomputer-based 2D general-purpose drafting system. Chapter 13 contains a complete discussion of the ways to go about evaluating and then selecting a CAD system; and the Appendix provides an extensive checklist of minimum features and functions that a good low-cost CAD system should have to be useful.

3

FUNCTIONS OF A 2D GENERAL-PURPOSE CAD SYSTEM

Figure 3.1 An APPLE-based CAD system with CPU, graphics screen, two disk drives, a digitizer for input, and an inexpensive plotter for producing hard copies. (Courtesy of APPLE Computers and T&W Systems.)

A low-cost 2D general-purpose drafting system creates a symbolic drawing using a digitizer or joystick for input, a graphics screen for review and editing of the drawing, and an X-Y plotter for placing the drawing on bond, Mylar, or vellum. The system employs a graphics CRT which is used to create a rough sketch of the final drawing. The drawing can then be plotted on a high-quality, high-resolution output device, such as a plotter or a printer. The system can be used to build up groups (or files) of complex symbols from primitive objects and text. These objects can then be used to make a complex drawing.

BASIC COMPUTER CONCEPTS

Initially, a computer receives information as "input" from a keyboard or digitizer. It then processes that information and displays the result (i.e., "output") on the graphics screen or plotter. For the 2D CAD drafting system, the computer uses data (the input) provided by the designer to produce (by processing the data) a drawing (the output).

THE 2D SYSTEM HARDWARE

The 2D drafting system consists of the following five hardware components:

1. The central processor (the "brains") and the keyboard
2. The floppy disk drives or, in more sophisticated systems, the "hard" Winchester-type disk drives (mass storage areas)
3. The display monitor (The CRT or screen, Figure 3.2)
4. The digitizer or joystick (input device, Figures 3.3 and 3.4)
5. The plotter (output device, Figure 3.5)

Figure 3.2 A CRT display monitor. (Courtesy of APPLE Computers and T&W Systems.)

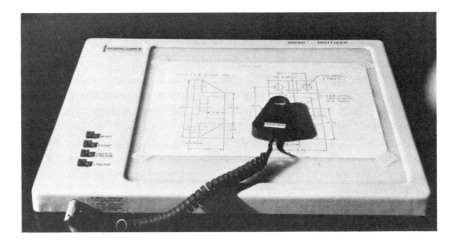

Figure 3.3 A digitizer for input. (Reprinted from CAD-TUTOR, by permission of Cadventures Unlimited.)

Figure 3.4 A joystick for input. (Reprinted from CAD-TUTOR, by permission of Cadventures Unlimited.)

Figure 3.5 An 8-pen "B" size plotter. (Reprinted from CAD-TUTOR, by permission of Cadventures Unlimited.)

The display monitor is used to display both text and graphics. (Often, 2D systems use two monitors, one to display text and a more powerful high-resolution monitor to display the graphics.) The keyboard and the digitizer are used for entering information into the computer. After input, the information is stored on a removable recording media such as a "floppy" disk (or diskette). A floppy disk is a thin piece of Mylar encased in a protective jacket with a surface that can be magnetized to store data. Many of the more powerful microcomputer systems now offer the storage option of a "hard" (or "fixed") disk. Hard disks have greatly enhanced storage capacity and provide the system user with greater flexibility. The plotter is used to provide a "hard copy"—on paper, Mylar, or vellum—of the drawing entered into the computer.

INPUT TERMINOLOGY

The 2D program receives its graphic input from the digitizer (sometimes called the graphics tablet) or from the joystick, in addition to the computer keyboard. When using the digitizer, the word "cursor" refers to the device that the designer moves across the digitizer, and the words "cursor button" refer to the button that is located on top of the cursor. When using the joystick, the word "cursor" refers to the handle of the joystick, and the words "cursor button" refer to the button on the joystick.

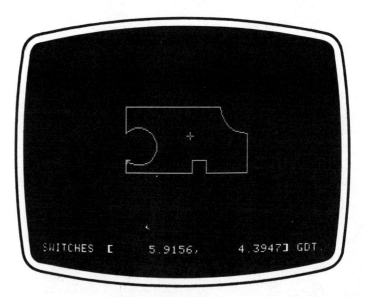

Figure 3.6 A graphic display with the small crosshair identifying the position of the cursor on the screen.

Often a stylus that looks like a typical ballpoint pen attached to the digitizer is used as an input device. The stylus enables the operator to "pick" objects or symbols directly off the digitizer pad; and to "pick" commands off the screen by pointing to them.

Whether the designer is using a digitizer pad, joystick, or stylus the cursor's position on the screen is represented by a small crosshair. As the designer moves the input device cursor, the crosshairs move across the screen. Throughout this book the word "cursor" is used to refer to both the input device cursor and the small crosshairs on the screen.

The cursor is used to create various graphic objects, such as lines, circles, and rectangles. It is also used to graphically manipulate the entire picture or any part thereof. These actions are performed by selecting the appropriate 2D function, positioning the cursor, and then pressing the cursor button. The program "beeps" or sounds a bell or in some other way acknowledges acceptance of the designer's action. Throughout this book the word "accept" refers to the pressing of the cursor button.

MENUS

2D programs are organized around a collection of many "menus." A menu is simply a list of the possible options available to perform a set of related operations. Each menu has a specific purpose. For example, a FILER Menu such as the one illustrated in Figure 3.7 allows the designer to enter com-

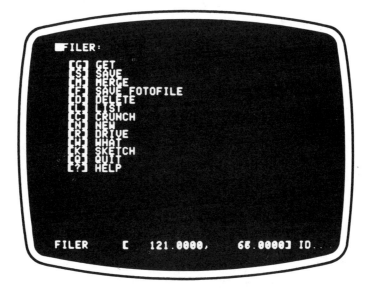

Figure 3.7 The FILER Menu.

mands that manipulate the graphics files, and an OUTPUT Menu allows the designer to send graphics drawings to an output device such as a plotter. Menus can be displayed on a separate screen or along the side or at the bottom of the graphics screen. CAD systems that utilize two screens have the menu displayed on one and the graphics on the other. Menus appear differently in the various microcomputer-based 2D systems. In the sample program that is featured in this book, each menu begins with a capitalized title followed by a list of the choices available. To select a choice listed on any menu, the designer simply presses the key for the letter between the brackets. Other systems might require the designer to enter a set of numbers or to type in a word or a word set.

BASIC 2D SYSTEM ORGANIZATION

In addition to menus, all 2D general-purpose drafting systems have certain features or functions in common. Some may provide more functions than others, or may run on larger or more powerful computer systems. But all systems utilize a computer, computer programs (or software), and libraries of symbols to increase productivity in the drafting process.

The system referred to in much of this book is a composite of several closely related 2D general-purpose drafting systems of moderate capability. It is used as a model to illustrate the features, functions, operations, and range of capabilities characteristic of most 2D systems.

Every 2D system works on the premise that:

- A designer creates objects.
- The objects can be combined (into groups, symbols, or cells) to represent more complex pictorials.
- In addition to being represented on the graphics screen, these pictorials can be drawn on high-resolution plotters or "dumped" to printers with graphics compatibility.
- Saved drawings can then be retrieved and modified or used as the foundation for still more complex drawings.

The following are features common to most 2D systems, although the terminology employed to describe them may vary or they may appear in a slightly different format in each. Wherever they occur, such variations are noted in the text.

OBJECTS

A 2D program allows the user to create a drawing using basic building blocks called objects. In other systems objects may be referred to as primitives (or primitive objects), entities (usually several primitives in combination), or features. These basic primitive objects may consist of lines, rectangles, regular polygons, circles, ellipses, circular arcs, dimensions, and text. Some systems may have more elaborate objects, such as Bezier (French) curves, available to the designer.

PROPERTIES

As primitive objects are created by the designer they are given several attributes or properties by the 2D CAD program that determine, in part, what they look like on the screen or when plotted. An example of a common property is the type of linestyle (e.g., solid, dotted, dot-dash, etc.) that the program uses to plot the object. In most systems these default property values can be changed by the designer at any time and thereby provide the designer with maximum flexibility or control over how the drawing looks on the screen when plotted.

GROUPS

A 2D program also allows the designer to manipulate the objects in the form of "groups" (referred to as "cells" in some systems). A group is an arbitrary user-defined collection of objects. The only relationship that objects in a group may have to one another is that they share the same group name. Groups provide a convenient way of manipulating large components of a complex drawing. They allow the designer to move, copy, rotate, or otherwise manipulate large numbers of objects at one time. In most programs, however, you cannot "nest" groups; that is, an object cannot be a part of more than one group at one time.

SYMBOLS

Symbols are related to groups, but are used differently in most CAD programs. A symbol is a collection of primitive objects, such as lines, arcs, or circles, and even other previously defined symbols, that make up a complex but *specific* figure. Once defined, the symbol itself is treated as a primitive object within the 2D program, and all operations that are valid for other primitive objects are valid for symbols.

The components of a symbol cannot be modified individually. A symbol is always treated as an indivisible unit. Even though a symbol may look complicated, the 2D program treats it just like a line or any other primitive object. This is in contrast to groups, which are also collections of primitive objects, but all of the objects in the group can be individually modified.

Typically, a symbol is a figure that is specific to a particular application and is used over and over again. For example, the standardized figure for a resistor is used frequently in electrical schematics and is therefore a good candidate for a symbol. Designers use symbols to build comprehensive application-specific symbol libraries.

LEVELS

Most 2D programs allow the designer to manipulate these objects on many different levels (or layers). A level or layer can be thought of as one sheet of clear plastic overlay. All objects identified with the same level number can be thought of as being on the same plastic sheet. "Turning on" one level is like viewing only one sheet. Turning on a second level or layer can be thought of as placing a second clear plastic sheet on top of the first—the objects on both sheets (or levels) can now be seen. Turning off a level is like removing the corresponding plastic sheets.

Levels are most useful if the designer can group together related objects in a common level. For example, if an architect is working on a floor plan, the architect might want to put the basic floor plan on level 1, the electrical plan on level 2, the furniture on level 3, and so on. Then the architect can work with any combination of the levels at any time.

Although they provide similar efficiencies of organization and speed, it is important for the designer to remember that groups and levels are completely independent of each other. For example, turning off a particular level could affect objects in several different groups. Similarly, moving a group could move objects in several different levels.

WINDOWS

Most 2D programs allow the designer to focus on a particular portion of a complex drawing, or correspondingly, to draw back away from a detail in order to get an overall view of the whole drawing. In some 2D systems this alteration of perspective or view is achieved through a dynamic function called "zoom," with movement on the horizontal plane (i.e., side to side) known as "pan." In other systems pan and zoom are combined in a single feature called "windowing."

The sample 2D program allows the user to view the drawing through any user-defined viewing "window." To better understand this notion of a window, imagine that the graphics screen of the computer is a movable window through which a drawing is viewed. The CAD program allows the designer to move the viewing window and thus to look at different portions of the drawing (i.e., pan). In addition, the designer can shrink the window and focus on a tiny detail (i.e., zoom in), or expand the window and look at a large, overall view (i.e., zoom out).

Shrinking the window and focusing on a detail sometimes gives the appearance of moving in toward an object. The closer you get, the larger the object or objects appear. It is because of the way that this feature is perceived that it is known as "zoom in" on some systems. The complementary situation, produced by moving away from an object, or stepping back from it, is known as "zoom out." Think of a camera with a zoom lens and the visual images produced by it.

REAL-WORLD COORDINATES

All numeric input to the 2D program is in terms of real-world X, Y coordinates. For example, objects are placed on the screen by the designer using real-world coordinates. Features that help to define the screen or to add precision to the screen (e.g., grids and windows) are also defined using real-world coordinates.

In a 2D CAD system, real-world coordinates reflect the actual size of the object the designer is drawing. For example, if an architect was designing a room addition for a house, all of the information about the room addition would be entered at actual scale. For example, if the width of the room is 20 feet, the lines making up the walls of the room would be entered as 20 feet. This process makes it easier on the designer, since he or she does not have to worry about scaling operations. All scaling is done when a plotted hardcopy is needed.

INPUT MODES

Each 2D CAD system has a variety of input modes: means by which a designer may specify X, Y coordinates while creating or manipulating a drawing. The designer has total freedom over which of the input modes to use, and it is usually quite easy to switch from one mode to another.

In some systems the digitizer and the joystick modes are referred to as "free" modes. In these modes, moving the digitizer or joystick moves the cursor on the screen. The arm movements of the designer are translated by the program into cursor movements on the screen. As the digitizer cursor or joystick handle is moved, objects can be made to follow the cursor on the screen. This feature makes the "free" modes a very visual form of input because you can always see, and immediately change, the relationship between the various pieces of the drawing. The free modes make use of the

Figure 3.8 Using a digitizer for input. (Reprinted from CAD-TUTOR, by permission of Cadventures Unlimited.)

Figure 3.9 Using a joystick for input. (Reprinted from CAD-TUTOR, by permission of Cadventures Unlimited.)

snap feature to ensure precise placement of the cursor on the drawing surface.

The other three input modes available to the designer in the sample 2D program are collectively called the coordinate input mode. These three modes enable the designer to type in the actual coordinates using the keyboard. The coordinates can be defined in three ways: absolute, relative, or polar. The coordinate input mode does not use the snap feature. If the snap mode is turned on, the coordinate input mode simply ignores it. The placement of objects in this mode, then, is a direct reflection of the coordinates inputted by the designer from the keyboard.

The absolute input mode is the easiest to understand. In this mode, the program requires the designer to input the absolute real-world X and Y coordinates. The program uses the coordinates provided by the designer to do its calculations and to identify the proper placement of an object.

In the relative mode the program requires that the designer provide

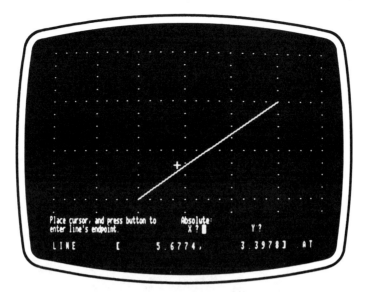

Figure 3.10 Using Absolute input to add a line.

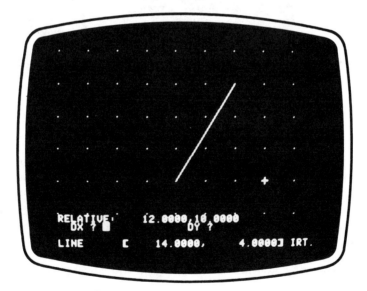

Figure 3.11 Using Relative input to add a line.

relative coordinates: that is, coordinates that are relative to a reference point on the drawing. Initially, this reference point is 0, 0. Thereafter, the reference point is set to the last X, Y coordinate defined. In some systems the reference point is reset back to 0, 0 when the program enters certain menu options.

In the sample program, if the MODIFY Menu option is selected in the relative mode, the reference point is initialized to the blinking object's "handle point." For example, a line's handle point is the first endpoint; a circle's handle point is its center point. Thereafter, the reference point is set to the last X, Y coordinate defined.

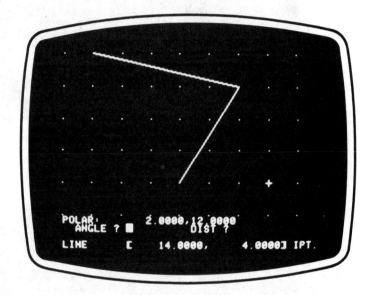

Figure 3.12 Using Polar input to add a line.

The third coordinate input mode in the sample program is the polar mode. This mode enables the designer to use a polar coordinate system. In this mode the program requires that the designer enter the angle and distance to the polar point. The angle is in degrees and is measured counterclockwise from the horizontal. The distance is in real-world units. The information entered through the polar mode is relative to the same reference point as described above for the relative mode.

The five modes can be used in any combination. For example, it is possible to define one endpoint of a line using the digitizer, switch to the relative mode, and then define the second endpoint of the line in relative coordinates. The flexibility provided by various modes and the ability to combine them as required can be very useful to the designer.

In the sample program, if the drawing's origin has been shifted, any coordinates entered using the digitizer, joystick, absolute, relative, or polar input modes are automatically shifted to be relative to the new origin.

INPUT FROM THE KEYBOARD

Much of the information that any 2D program needs is supplied by input from the digitizer or the joystick. However, it is sometimes necessary (or desirable) to type information in at the keyboard, in addition to using the keyboard for the three coordinate input modes. The types of information that may have to be entered at the keyboard fall into three classes: text characters, numbers (both real and integer), and answers to short questions. For example, depending on the system and the circumstances of the moment, the designer might be required to enter the names of groups, the floating-point values of grid spacing, or to answer "Y" or "N" to a question or prompt.

Text characters are entered by typing the appropriate characters on the keyboard. Single-character mistakes can usually be corrected by using either a backspace key (if there is one on the computer) or the left arrow key. Each time the left arrow key or the backspace key is hit, the last character is erased.

The RETURN (or ENTER, on some systems) key is usually used to terminate the entry of a string of text characters. Hitting the RETURN key tells the program that characters just entered from the keyboard are correct and that it should read and process them.

There are two types of numeric data that may be entered from the keyboard, integer numbers and real numbers. Both types of numbers are entered by typing the appropriate digits on the keyboard. Any "illegal" keys that might be typed are simply ignored by the program. The backspace or left arrow key is used for correcting single-digit mistakes. Each time the backspace or left arrow key is hit, the last digit is erased.

The ESCAPE key is used to abort the operation that required the entry of numeric data. In general, the designer should use the ESCAPE key if an option is selected accidentally and the designer wants to leave without doing anything.

The RETURN (or ENTER) key is used to terminate the entry of a number. Hitting the RETURN key tells the program that the number just typed in at the keyboard is correct and that it should read and process it.

The range of real integer numbers that can be typed in at the keyboard varies according to the CAD system and computer system being used. The sample CAD program that is featured in this book is limited to a range of −32767 to 32767, and does not accept any numbers outside this range. If a number outside this acceptable range is typed in, the program "beeps," erases the number just entered, and repositions the designer to enter another number.

The sample 2D program characteristically asks several questions that require a single one-character answer. These include yes or no questions or selecting a variety of options from the menus. All of these questions can be answered by responding correctly to the prompt, usually by typing a single character. In general, any question that can be answered by typing a single key does not need to be terminated with a RETURN; the program reads the response and processes it automatically.

FUNCTION KEYS

Most microcomputers are equipped with programmable function keys that enable the user to build in special general functions that are used repetitively. Similarly, most 2D CAD programs provide function keys to activate special system features. A unique thing about function keys in these 2D systems is that they can be used any time the cursor is in use. For example, in our model system, it is possible to turn the increment snap option on or off while performing a very complex operation.

Most of the function keys in 2D CAD systems provide immediate visual feedback, telling the designer that the function was activated successfully. Other function keys display a character near the coordinate dial on the status line at the bottom of the screen.

In the sample 2D program, operations that can be activated through the use of function keys include:

- Turning the coordinate dial on and off (turning the dial off speeds up the computer program's processing speed)
- Activating and deactivating the joystick and digitizer scaling function. (When the scaling function is activated, the designer restricts the movement of the cursor to one-fourth of the graphics screen presently accessible. This function may be repeated until the cursor is successfully placed on a precise screen dot location.)
- Switching among the three snap modes available: increment or grid snap, or free sketch.
- Switching the object tracking mode on and off.
- Switching among the three available screen display modes: full text, full graphics, or a screen consisting of one-sixth text and five-sixths graphics.
- Switching between a short and a long cursor.

Each of these function keys is described in greater detail below. Additional function keys that might be encountered in other 2D systems might enable the designer to shift between the shifted and nonshifted origin of a drawing, to inquire into and/or change any of the global properties of an object, to dump either the graphics or alpha portion of the screen to a graphics printer for quick copying, or to stop the computer from executing a particular operation.

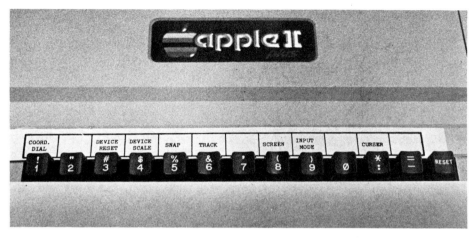

Figure 3.13 CAD function keys on the APPLE microcomputer. (Courtesy of APPLE Computers and T&W Systems.)

Figure 3.14 CAD function keys on the IBM-PC. (Reprinted from CAD-TUTOR, by permission of Cadventures Unlimited.)

THE COORDINATE DIAL

The coordinate dial, located in the lower right-hand corner of the text screen of the sample program, provides the designer with a continuous or dynamic readout of the cursor's real-world location. The X, Y coordinates displayed are always in terms of real-world coordinates. The sample program displays the seven most significant digits on the coordinate dial. The level of precision, of course, varies from system to system.

JOYSTICK SCALER OPERATION

Several 2D CAD systems support the use of a joystick as one of the graphics input devices. The joystick is a very economical input device, but its accuracy and operation does not match that of the digitizer. Because of the joystick's inherent lack of resolution, the sample system provides a special function key feature to help remedy this deficiency. This function key can be used to scale the joystick's movement down by a factor of one-fourth each time it is pressed. This means that the full swing of the joystick's handle is mapped into a smaller screen area, thus providing an increase in accuracy. A second function key is used to reset the joystick's movement to the full screen area. (This function key operation can also be used to enhance the precision of a digitizer pad.)

SNAP

It is sometimes difficult to precisely position an object using the cursor. This is due to the visual problem in placing the cursor exactly on a desired X, Y location. Even if the cursor is moved so that it does appear to be on the desired location, the limitations of the resolution of the graphics screen on most systems almost guarantees that it is not exactly where the designer wants it to be. It would seem that the only way to escape from this problem is to type in the coordinates using the coordinate input mode.

In order to address this problem in precision—or the lack thereof—a "snap" or "pick" feature has been implemented on many CAD systems. Snap and pick round the cursor and coordinate dial to an exact X, Y coordinate location. In the sample program featured in this book the state of the snap function is displayed on the screen near the coordinate dial. The sample program provides a special function key feature to cycle through the available snap modes.

Increment Snap: When increment snap (called "increment pick" on some systems) is turned on, the program rounds the position of the cursor and coordinate dial to the nearest user-defined increment. Thus the designer is able to obtain a precise X, Y position on the graphics screen. For example, if the user-defined increment is 0.125 inch, each point entered with the digitizer or joystick is automatically rounded to a multiple of 0.125 inch. In the sample program, an "I" is displayed near the coordinate dial on the text screen when increment snap is on.

Grid Snap: When grid snap (or grid pick) is turned on, the program rounds the position of the cursor and the coordinate dial to the X, Y value of the nearest grid intersection point. Thus the designer is able to obtain a precise X, Y position on the graphics screen. For example, if the grid spacing was set to 1 inch, each point entered with the digitizer or joystick is automatically rounded to a multiple of 1 inch. In the sample program, a "G" is displayed near the coordinate dial on the text screen when grid snap is on.

No Snap: The third snap mode in the sample program disables both increment and grid snap. When "no snap" (or free pick) is turned on, the cursor and coordinate dial use the best resolution possible. In the sample program a "." is displayed near the coordinate dial on the text screen when no snap is on.

If any of the coordinate input modes are being used, such as absolute, relative, or polar, the state of the snap function will not affect the values entered from the keyboard. That is, if increment snap is turned on and the absolute mode is being used, the absolute X, Y coordinate typed in from the keyboard is not rounded to the nearest user-defined increment.

OBJECT TRACKING

Normally, all drawing manipulation in 2D CAD systems is done interactively; that is, objects change size and orientation on the screen as the designer moves the cursor. This provides for very effective visual feedback. This interactive mode is usually called object tracking—or, in some systems, "attentioning" an object—since all graphic objects track or follow the movements of the cursor.

Although it aids in the placement of objects on the screen, one possible disadvantage of using object tracking (or attentioning) extensively is that it can slow down the response time of the program. In the sample program a special function key feature enables the designer to turn on the tracking mode, or to turn it off whenever increased speed is desired. When object tracking is on, a "T" is displayed on the text screen near the coordinate dial.

LONG AND SHORT CURSOR

The normal short cursor that appears on the screen in most 2D CAD programs looks like a small crosshair. Some programs also provide an alternate long cursor for use by the designer. The long cursor is a large crosshair that extends across the entire screen. In addition to being entirely equivalent in function to the short cursor, it provides a means for the exact visual alignment of objects. In the sample program a special function key feature enables the designer to switch between the long and the short cursor.

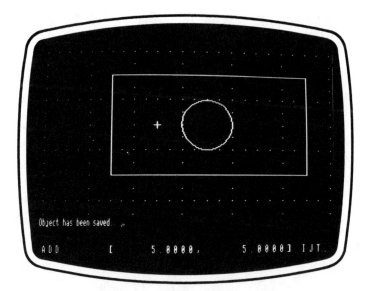

Figure 3.15   Using the short cursor.

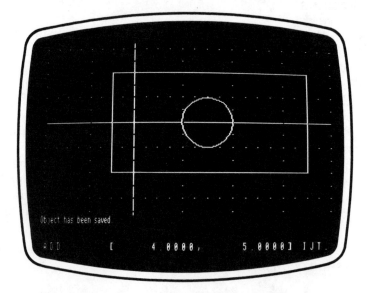

Figure 3.16   Using the long crosshairs for positioning.

SCREEN DISPLAY MODES

Some programs feature the option of various screen display modes—or even multiple screens. In the sample 2D program there are three alternative screen display modes. The first screen displays nothing but text. All of the program menus and messages are displayed on this screen. The second screen displays nothing but graphics. All of the drawing is displayed on this screen.

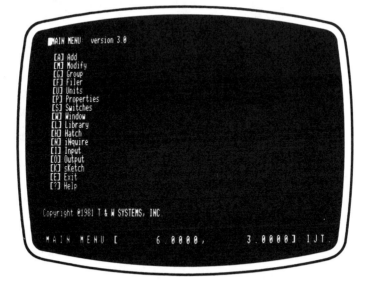

Figure 3.17 Full text screen mode.

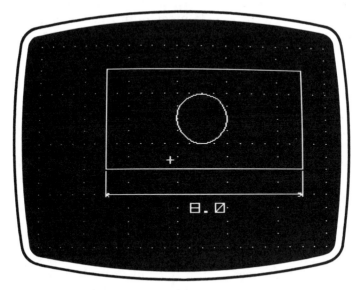

Figure 3.18 Full graphics screen mode.

The third screen displays both text and graphics. The upper five-sixths of the screen is dedicated to graphics and the lower one-sixth of the screen displays text. This third screen is very useful because the designer can view

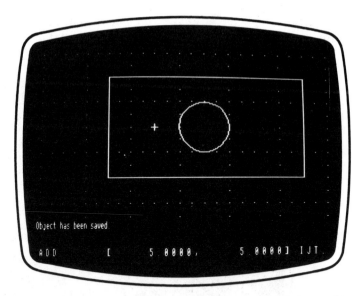

Figure 3.19 Split screen mode—one-sixth text, five-sixths graphics.

the drawing, coordinate dial, menu title, and various program messages simultaneously. The sample program provides a special function key feature that enables the designer to switch among the three screen display modes.

STATUS LINE

Most 2D systems with any degree of sophistication permit the designer to "enable"—turn on—several operations or program functions at one time. Each program must provide a way for the designer to maintain track of those operations or functions that have been enabled or activated. Occasionally, activated functions or operations are displayed by a 2D program as inverted characters (e.g., black on white instead of the program's normal white characters on a black screen). But often this is done visually directly on the graphics screen through a "status line" feature.

In the sample 2D program the status line is displayed as the bottom line of the graphics screen when the text or split screen mode is enabled. Typicallly, the screen with menu and status line appears as shown in Figure 3.20. In this example, the first field displays the current menu title—ADD Menu. If another menu option is selected, that menu option (e.g., FILER, MODIFY, GROUP, etc.) is displayed in this field.

The second field displays the current real world X, Y position of the graphics input device (0.0000, 8.2401), such as the digitizer or the joy-

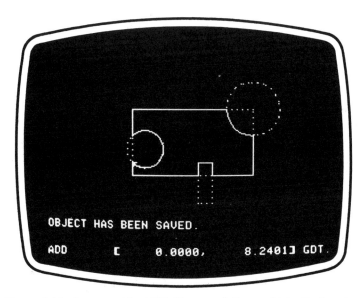

Figure 3.20 Screen with ADD Menu and Status Line displayed.

stick, on a coordinate dial. When the position of the joystick or digitizer is changed, the program immediately updates the dial with the new X, Y position.

The third field displays the state of the snap mode. As noted above, in addition to having a free sketch mode, most programs have features that cause the cursor and the object or group being drawn to "snap" to the nearest increment or closest grid intersection point. In some systems increment and grid snap are referred to as increment and grid "pick." In the sample program, an "I" displayed in this third field on the status line indicates that increment snap is on. "G" indicates that grid snap is on. A "." in this field indicates that neither increment nor grid snap has been enabled and that the program is operating under free sketch. In the illustration above, grid snap is turned on.

The fourth field displays the state of the coordinate input mode. Thus all X, Y coordinate data must be entered in the appropriate manner. In this example a "D" in this field indicates that digitizer input is required. A "J" requires joystick input. Absolute, relative, or polar keyboard input are required when "A", "R," or "P," respectively, are displayed.

The fifth field displays the state of the "object tracking" mode. When the object tracking mode is turned on—a "T" is displayed in this field— objects are shown by the program to interactively follow the path of the digitizer as described by the designer. In certain 2D systems, object tracking is referred to as "attentioning" an object.

The sixth field displays the state of the drawing's origin. If an "O" is displayed in this field, it indicates that the drawing's origin has been shifted from its original X, Y location.

HANDLE POINTS

When objects are initially created by a designer, they must be located on the graphics screen using some predetermined reference point on or within the object. Later, when an object needs to be moved or rotated the cursor needs some definite reference point to latch onto so that the object's new location or orientation can be identified correctly. In 2D programs these reference points are known as handle points. This handle point is the point by which the cursor holds onto an object or group while it is being manipulated. For example, when a group is moved, it is pulled across the screen by its handle point.

Handle points can be predetermined by the program or defined by the user, or both. In the sample 2D program, all objects are given a default handle point when they are created. Several objects are also given a secondary handle point. These handle points are defined in Table 3.1.

TABLE 3.1   OBJECT HANDLE POINTS

Object	Initial handle point	Secondary handle point
Line	First endpoint defined	Second endpoint defined
Rectangle	First corner defined	Second corner defined
Polygon	Center	None
Circle	Center	None
Ellipse	Center	None
Arc	Center	None
Bezier curve	First endpoint defined	Second endpoint defined
Text	Lower left-hand corner	Upper right-hand corner

TEMPLATE OBJECTS

Most low-cost 2D programs do not support hidden-line removal or the drawing of partial objects, but most do support the use of "template" (sometimes known as "guide") objects. For example, suppose that there is a rectangle on the screen and a circle is to be placed in such a way that it appears as if it is in front of the rectangle. Once the circle is placed, the rectangle is still completely visible. A hidden-line removal feature would remove from the screen the sections of the rectangle that are behind the circle. A partial object's feature would allow the designer to modify the rectangle manually so that sections are removed from view. Most 2D programs use template objects to solve this problem.

The sample 2D program uses a unique approach to do hidden-line removal by putting some of the responsibility on the designer. In order to

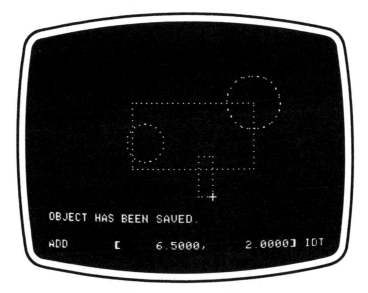

Figure 3.21 Using Template objects as a guide.

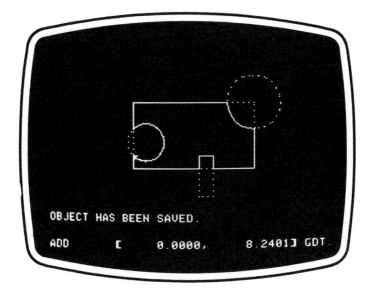

Figure 3.22 Drawing solid lines and objects over Template objects.

give the appearance of a circle in front of a rectangle, the program allows the designer to create template objects. A template is really nothing but a standard 2D object given the property of a template object. When defined as a template the object is drawn with dots. These dots are the key to making the section of the rectangle behind the circle disappear from the screen.

The solution to this problem is now really quite simple. First, the designer should draw a template rectangle on the screen, and then draw a nontemplate circle in front of it. The rectangle is still fully visible on the screen, but in the form of dots. Lines are then used to trace over the dotted sections of the rectangle that are to remain on the screen. When the designer is satisfied with the drawing, he or she should delete the template rectangle from the drawing. The drawing now shows a circle in front of a rectangle with the hidden sections of the rectangle removed.

THE WORKFILE

2D programs always keep a copy of the drawing on which the designer is currently working in a "workfile." A workfile is just a temporary file, usually on diskette, that is used by the 2D program only to store this copy of a drawing. Working on the data within this workfile has no effect on any drawings saved in other files on the diskette.

The designer is never working on the original copy of the drawing, unless the drawing is a completely new one. Instead, it is always the copy stored in the workfile that is being added to or revised. By always working on a copy the designer is assured that any drastic mistakes affect only the copy and not the original.

Drawings that a designer is working on should periodically be saved to diskette, or to the hard disk. Saving a drawing causes the program to write the current copy of the drawing in the workfile to a permanent file. The practice of periodically saving a drawing provides the designer with permanent backup in case the workfile has been manipulated to the point where it is so complex—with errors, as well—that further work, or even the correction of what has already taken place, is impossible. By saving the workfile periodically, and then recovering the contents of the drawing before a major disaster has occurred, the designer is spared the time-consuming task of having to go back and undo mistakes or having to start again from scratch. In certain 2D systems the designer can set the program to provide regular periodic prompts for saving new objects created or modifications made to existing objects or groups in a workfile.

The workfile should also be saved as a permanent disk file whenever a drawing is completed. This permanent disk file can later be retrieved and put back into the current workfile. Files saved to a hard disk can similarly be retrieved for use at a later time. Remember, the designer should always save the workfile to diskette before leaving the 2D program.

RECOVERY

One great fear of all designers using CAD programs is that they will be unable to retrieve their work at a later time, and that time and effort will have been wasted. Most 2D CAD programs provide the designer with rather complete "recovery" features. The recovery feature makes it virtually impossible for the designer to lose any work, even if the computer should lose power or for some other reason "crash" or "go down." Except under the most unusual circumstances, the most that will be lost after a machine failure is the result of the very last program operation.

The sample 2D program accomplishes this recovery by constantly updating the workfile that is stored on the designer's diskette, or on the hard disk. As the designer makes additions or changes to a drawing, those additions and changes are recorded in the workfile. As noted above, this updating function can be done automatically by the program or at intervals prescribed by the designer.

In addition to saving a copy of the drawing in the workfile, the program also saves the current values of all the defaults and switches—the global characteristics of the drawing—all windows that have been saved, all sets of special plotter specifications (plot specs) that have been defined, and the current viewing window. This means that the workfile contains all the current information needed to restore the working environment completely should an unexpected interruption occur.

TUTORIALS AND HELP FEATURES

Many microcomputer systems and software programs on the market today have tutorials built into them for the first time and even for the more advanced user. Tutorials are usually designed to help the user get "up and running" with the microcomputer system or the applications software program as quickly as possible.

Still other software programs have an interactive "help" feature that enables the program user to access assistance on the use of a particular program feature or function from any point within the program. Many CAD programs have similar tutorial and/or help features built into them.

The "Help" option built into the sample 2D program provides the designer with helpful information about the 2D drafting system as he or she

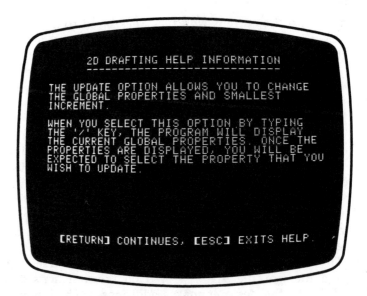

Figure 3.23 Screen illustrating the built-in Help feature of the sample system featured in the text.

is working on the drawing. It is like a handy reference card that briefly explains each available menu choice. In this program the Help option can be selected from any menu that displays the option. To access the Help option, the designer need only enter the "?" character.

4

STARTING A DRAWING: 2D COMMANDS, UNITS, PROPERTIES, AND SWITCHES

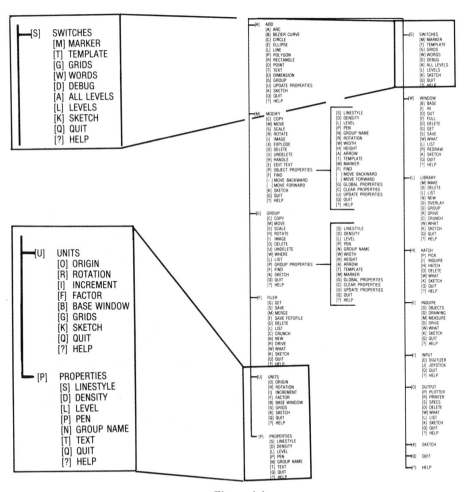

Figure 4.1

USE OF 2D COMMANDS

Most 2D CAD programs are organized around menus which lead the designer into the functions and features of the system. In most instances, program menus are structured in a branching arrangement and a main or central menu branches out to a series of subordinate secondary and even tertiary menus.

THE MAIN MENU

In the sample 2D program featured in this text the program MAIN Menu is displayed automatically when the program is "booted up" or initialized. It can be reached from any point in the program by simply using "Q" for "Quit," to exit any program function or operation. In the sample program, each "Q" entered moves the program from the present menu to the menu one level higher. The program does not "nest" menus more than three deep, so returning to the MAIN Menu from the lowest level takes minimal time.

In the sample program the MAIN Menu appears as shown in Figure 4.2. Note that each subordinate menu—ADD, MODIFY, GROUP, and so on—is coded by a single letter representation. Other systems use two-digit numbers or word sets to initialize menu functions.

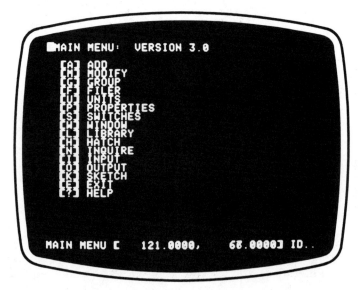

Figure 4.2 Typical 2D general purpose CAD system Main Menu with options displayed.

The MAIN Menu has one specific purpose, and that is to provide the designer with a means to reach the specific 2D functions needed to create drawings. This menu can be thought of as the main hallway containing many doors, or as the hub of a wheel containing many spokes. Each door in the hallway, or spoke in the wheel, performs no specific task except to lead into a room filled with options that actually do perform tasks.

For example, the ADD Menu is a "door" on the MAIN Menu. Simply choosing the ADD option from the MAIN Menu does not, however, enable the designer to add objects. Rather, the designer must select from the several options provided within the ADD Menu when it is displayed.

When a designer starts a drawing using a 2D system, the designer must define several parameters. Although these parameters vary from system to system, they usually include the drawing size, the grid definition, the size of the smallest increment, and information related to the drawing origin and any shifted origin. Regardless of the system being used, these parameters must be defined before any work is begun. In the sample 2D program featured in this and later chapters, the UNITS option allows the designer to examine and/or change the various default parameters that are used throughout the program.

When any object is created—a line, circle, polygon, and so on—it is given several initial properties, or attributes, that determine what it looks like when it is drawn. Examples of a property are the standard size of a text character, the plotter linestyle, and the group name. For the most efficient drawing, properties of an object should be determined before an object is created. A number of functions do exist in most 2D programs that allow the designer to modify properties once established—for example, see the section on PROPERTIES below—but this is time consuming and detracts from the productivity that can be gained if drawings are carefully planned in advance.

Completed drawings are then saved to a diskette, or onto a hard disk. Proper file management—the conventions related to saving, storing, retrieving, and archiving drawings—then become important considerations for the designer. Options in the FILER Menu of the sample program, and some guidelines for proper file management of a 2D CAD system are presented in Chapter 6.

UNITS

When the designer selects the UNITS option, the screen displays the current working parameters, as shown in Figure 4.3. When the designer selects an option in the UNITS Menu, the program positions the cursor to change the value of that parameter. The designer enters the numeric value desired and

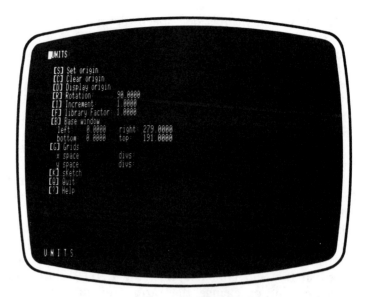

Figure 4.3 The UNITS Menu.

then hits the RETURN key. The new value is then displayed on the screen. In the sample program, hitting the ESCAPE key aborts the current option and returns the program to the UNITS Menu.

Some systems automatically designate a system of units such as feet, inches, or centimeters. Other programs allow the designer the flexibility of determining those units and require that the designer annotate each drawing so that it is understood that all units are in feet, inches, or centimeters.

ORIGIN

Most programs allow the designer to establish a particular X, Y location for each drawing and viewing window, and then to change that origin to meet particular needs.

The Set Origin option in the UNITS Menu of the sample program enables the designer to shift the base window's origin. When this option is selected, the designer moves the cursor to the position on the screen that he or she wants to be the new origin and accepts it by pressing the cursor button. If the origin was successfully changed, the program displays an "O" near the coordinate dial in the lower right-hand corner of the text screen. When the origin is shifted, all digitizer, joystick, and keyboard coordinates are converted to the shifted system.

The Clear Origin option in the UNITS Menu enables the designer to

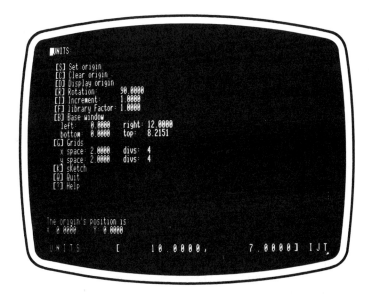

Figure 4.4 The Origin submenu of the UNITS Menu.

quickly reset the shift of the base window's origin to the X, Y coordinates 0, 0. When this option is selected, the program displays "." near the coordinate dial in the lower right-hand corner of the text screen.

The Display Origin option in the UNITS Menu enables the designer to verify the current shift of the base window's origin. When this option is selected, the program displays the current X, Y position of the origin.

ROTATION

The Rotation option in the UNITS Menu allows the designer to define the incremental rotation that is used throughout the sample 2D program.

There are several areas in the program that use this rotation feature. One of these is the Text option of the ADD Menu. Once text has been entered, the program allows the designer to rotate the text by an incremental amount. When this option is selected, the program prompts the designer to enter the new incremental amount, 0 through 360 degrees.

This incremental rotation then defines a specific amount of degrees that will be used in an operation allowing rotation. For example, if the rotation was defined to be 45 degrees, then each time the text is rotated, the program adds 45 degrees to its rotation. The text starts at 0 degrees, then moves on to 45 degrees, then 90 degrees, and so on.

INCREMENT

Most 2D programs allow the designer some control over designating the working units of a drawing. The Increment option in the UNITS Menu allows the designer to change the global default value of the smallest increment. The smallest increment defines the smallest incremental unit of precision that the program maintains for a given drawing—or portion of a drawing. It is used primarily with such functions as grid and increment snap (or pick), as an aid in the precise placement of objects. The initial default value of the smallest increment is usually set to 1.0. When this option is selected, the program requires that the designer enter the new value of the smallest increment.

When the increment snap function is on, the program automatically rounds the cursor to a multiple of the smallest increment. The designer is able to increase the resolution of the drawing on the screen by changing the smallest increment that the program will maintain at any time. When using increment snap, the cursor and coordinate dial may not always reflect every multiple of the smallest increment. If this is the case, the use of the WINDOW function should produce finer resolution and thereby allow more precise placement of the cursor.

LIBRARY FACTOR

Many CAD systems permit the designer to build complex groups that are used as standard symbols in drawings. These symbols are then organized into and stored as symbol libraries. Symbols can be recalled from their libraries and quickly added to new drawings, contributing directly to the productivity gains achieved in CAD systems.

The Library Factor option in the UNITS Menu enables the designer to change the global default value of the library scaling factor. This option is used when a designer is placing symbols on a drawing from a library. When the Library Factor option is selected in the sample 2D program, the program prompts the designer to enter the new value of the library scaling factor.

This option is especially useful if the designer wishes to scale all of the symbols in a library by a constant amount. For example, suppose that all the symbols in a particular library were originally defined in feet. The designer is now working on a drawing using metric units and requires the symbols to be redefined in meters. In this instance, the designer simply changes the library scaling factor to 0.3048 (the feet-to-meters conversion factor) and all of the symbols within the library are automatically scaled to meters when placed on the drawing.

BASE WINDOW

The Base Window option in the UNITS Menu allows the designer to set or change the basic coordinate system (or base window) that is used for a drawing. The base window is an arbitrary, user-definable, screen coordinate system that is meant to be used as the basic coordinate system while creating a drawing. What is the "best" base window is a matter of preference. For example, if an architect is drawing a floor plan that is 30 feet by 40 feet, the architect would probably want to set the base window to something slightly greater than 30 feet by 40 feet. Or if the drawing once completed is to be plotted on a 15-inch by 20-inch surface, the best base window to set might be one 15 by 20.

GRIDS

Each 2D program has a number of features built into it to aid the designer in the proper and precise location and placement of objects in a drawing—overcoming some of the limitations of the resolution of the graphics screen. So far, the use of function keys for turning on object tracking (attentioning), increment and grid snap (pick), and the digitizer and joystick scaler have been presented and their uses discussed.

The Grids option in the UNITS Menu enables the designer to place an X-Y grid on the screen that also aids in the precise alignment of graphic objects. Although grids might be created in different ways in different 2D programs, their function is the same in each.

When the Grids option is selected, the program positions the designer to enter the X-axis spacing. This spacing is defined in terms of real-world coordinates—and in relationship to the rest of the drawing and the base window—and represents the perpendicular distance between the vertical grid lines.

The program then prompts the designer to enter the number of divisions. The divisions field defines the number of units to be marked off on the X and Y axes between the grid intersections. For example, if the X-axis spacing entered is 1.0 and the grid divisions entered is 10.0, the program marks off 10 equal units between the grid intersections by plotting a dot every 1/10 of a grid spacing.

The Y-axis grid spacings and divisions are entered and plotted in the same way. Both X- and Y-axis grid lines must be specified by the designer.

There is always a dot plotted for each grid intersection. If the distance between the grid intersections becomes too small, the program does not display the divisions of the grid lines.

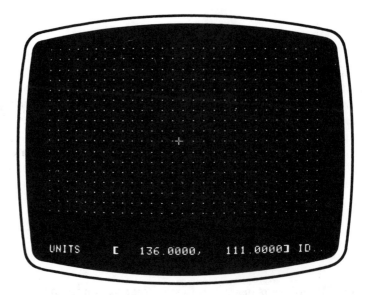

Figure 4.5 Establishing a grid pattern from the UNITS Menu.

A simple way to remove the grid system is to select the Grids option and just hit the RETURN key when defining the X-axis spacing. This action erases the old grids and, since new ones have not been defined, no new grids are displayed. In the sample 2D program, the Grids option can be turned off temporarily by using the appropriate option in the SWITCHES Menu.

SKETCH

Sometimes selective erasures, moves, and the like leave imperfect images on the screen. Each 2D program has a means to redraw or resketch the graphics screen so that these imperfections are removed.

The Sketch option in the sample program enables the designer to clear the graphics screen and to redraw the picture eliminating any imperfections in the graphics image that might have appeared while modifying the screen image.

Sketch is available as an option from the MAIN Menu as well as an option from within most of the other menus of the sample program.

QUIT

The Quit option performs the same function in each of the menus (and submenus) in the sample 2D program. When the Quit option is selected by the designer, the 2D program returns to the next highest menu—and, ultimately, if Quit is hit several times in succession, to the MAIN Menu.

HELP

As noted earlier, many software programs now contain built-in tutorials that assist the user with the operation of program features and functions.

In the sample 2D program the Help option provides the designer with helpful information about the 2D drafting system as the designer is working on a drawing. It is like a handy reference card that briefly explains each available menu choice.

When the Help option is selected, the program prompts the designer to identify the item on the menu for which assistance is required. Once the designer selects an option, the program then displays a brief, yet correct, explanation of the option that was selected. The explanation may be several screens in length. The designer can exit the explanation and return to the appropriate menu at any time by simply hitting the ESCAPE key.

Although the Help or tutorial feature may function differently in various CAD programs, the purpose or intent of such a feature is always the same.

PROPERTIES

The options in the PROPERTIES Menu enable the designer to examine and/or change the various initial properties that are used throughout the program. Figure 4.6 shows the PROPERTIES Menu as it looks in the sample 2D program. When the PROPERTIES Menu is selected, the program lists all of the available properties together with their current values on the

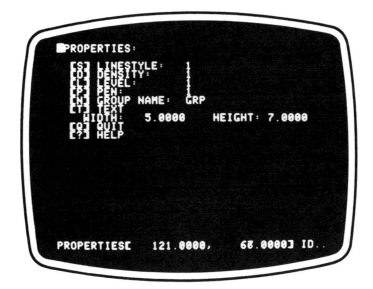

Figure 4.6 The PROPERTIES Menu.

left-hand side of the screen. The program positions the cursor so that the designer can change the current value of a property. Depending on the particular property, the designer enters either numbers or characters. As new values are entered, they are displayed on the screen. ESCAPE is used to abort the option with no change to the current properties.

LINESTYLE

A linestyle is a repetitive pattern that is produced by a plotter when it plots a line. The most typical linestyle is a solid line. Other linestyles available include dotted, dashed, and dotted-dashed lines. The linestyle option is really meaningful only if the plotter being used is capable of generating a variety of linestyles and the commands entered must correspond to the linestyle code of the plotter.

Not all 2D programs support a variety of linestyles on the graphics screen. In these programs all lines may be shown as solid lines on the graphics screen. Occasionally, dotted as well as solid lines may be displayed on the graphics screen. But in all instances, all lines appear with the correct linestyle when plotted.

When the Linestyle option in the PROPERTIES Menu is selected, the program requires the designer to enter the new linestyle value. The new default linestyle—that is, the type of linestyle the program will subsequently use for all lines and objects unless otherwise indicated—is then displayed on the screen.

DENSITY

The density of a line is a reflection of the apparent width and/or darkness of a line. If a multipen plotter is being used, line density can be achieved by varying the thickness of the plotter pens and coding density by using the Pen option (described below) in the PROPERTIES Menu. More often, though, variable line density is achieved by varying the number of times a plotter overstrikes an object. The greater the density, the darker the objects.

When this option is selected, the program positions the designer to enter the new density values. The new density value is then displayed on the screen.

LEVEL

A level can be thought of as a sheet of clear plastic overlay. All objects with the same level number are on the same plastic sheet. Turning on one level is like viewing only one sheet. Turning on a second level is like placing a second plastic sheet on top of the first—objects on both sheets (or levels, or overlays) can now be seen. Turning off a level is like removing the corresponding plastic sheet.

All 2D programs allow the designer to work with multiple levels—sometimes as many as 256. Usually, any combination of these levels can be used at any time. The Level option in the PROPERTIES Menu of the sample 2D program allows the designer to establish the level property for a given object or for an entire drawing.

When this option is selected, the program positions the designer to enter the new level value; the default level value is always 1. The new level number is then displayed on the screen.

PEN

As noted above, many plotters have more than one pen available at a time. These pens may be of different colors or different line weights (i.e., density). The pen numbers used in the program correspond to the different pen stalls of the plotter. For example, a pen number of 4 causes the plotter to use the pen that is in stall number 4. Of course, this option is meaningful only for those systems that provide plotters that support multiple pens.

When the Pen option in the PROPERTIES Menu is selected, the program requires the designer to enter the new pen number. The initial value of the default plotter pen number is 1. When a new value is entered, it is displayed on the screen.

GROUP NAME

Every 2D system provides features that enable the designer to assemble complex drawings. In most systems, as objects are created they are put into groups—groups that are defined by the designer.

In the sample 2D program the group names are entered or changed by using the Group Name option in the PROPERTIES Menu. Changing the name changes the group that objects are placed into when they are created. Any group function that is selected operates on *all* objects with the same group name.

TEXT

Just as the base window must be defined by the designer when starting a drawing, it is also necessary to define text width and height. Both text width and height are in terms of real-world coordinates. If the default base window in the sample 2D program is used (with screen width at 279 and height at 209), the program automatically sets text width to 5.0 and text height to 7.0.

The Text option in the PROPERTIES Menu allows the designer to change the text properties that determine how text characters are displayed when they are first created. When this option is selected, the program positions the designer to enter the new width property, and then the new height property. The new values are then displayed on the screen.

Different 2D programs may handle establishing text width and height in a variety of ways. When the use of default values are not a built-in feature of a program, text width and height may have to be established each time text is created and added to a drawing.

COLOR

Many computers have the ability to display objects on the screen in color. The number of colors available to the designer may vary dramatically from 2 to 16, or even more. A Color option in a 2D CAD program enables the designer to change the screen color property.

Each screen color property corresponds to one of the colors that can be displayed on the screen. For example, a screen color of 3 may correspond to the color red. Of course, this option is meaningful only for those systems that support color screens.

SWITCHES

Many 2D programs contain a variety of functions that aid in the creation of objects and their proper placement in the drawing and on the screen. Other functions are used to turn on and off special features or objects or special functions of a program. In the sample 2D program the options in the SWITCHES Menu allow the designer to examine and/or change the various switches that are used throughout the 2D program. These switches control whether a particular 2D function is turned on or off.

In the sample 2D program the SWITCHES Menu appears as shown in Figure 4.7. When the SWITCHES Menu is selected, the program lists all of the available switches, together with their current values, on the left-hand side of the screen. When one of the options is selected, the program positions the designer to enter the new value of that switch. The only acceptable values of a switch are "Y" for yes (on) or "N" for no (off). Each time an option is selected by the designer the program positions him or her to enter either a yes ("Y") to turn a feature on, or a no ("N") to turn a feature off. When the new value is entered, it is displayed on the screen.

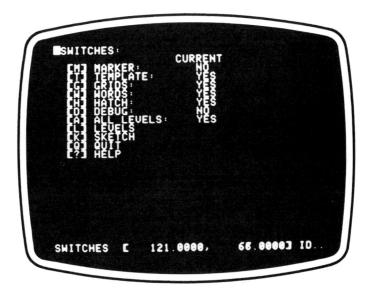

Figure 4.7 The SWITCHES Menu.

MARKER

The Marker option in the SWITCHES Menu allows the designer to change the global show center marker switch. If the center marker switch is set to "Yes," then all objects are drawn with a marker at their handle(s). If the switch is set to "No," the markers are not shown.

When this option is selected, the program positions the designer to enter either a "Y" (for yes) or an "N" (for no). Typing in any keys other than a "Y" or "N" causes the program to keep the old switch value.

TEMPLATE

The Template option in the SWITCHES Menu allows the designer to change the global show template switch. If the show template switch is set to "Yes," all template objects are drawn whenever the screen is redrawn. If the switch is set to "No," template objects are not shown.

GRIDS

The Grids option in the SWITCHES Menu allows the designer to change the global show grids switch. If the show grids switch is set to "Yes," the currently defined grid lines are shown whenever the screen is redrawn. If the switch is set to "No," the grid lines are not shown, even if they are currently defined in the UNITS Menu. Consequently, this option is a convenient way to turn grids on and off without having to redefine them each time.

WORDS

Since text is entered as any other graphics object, it has a way of slowing down the program whenever a screen has to be drawn or the designer has to search a file to find a particular object to modify.

The Words option in the SWITCHES Menu, and comparable options in other programs, allows the designer to change the global show words (or text) switch. If the show words switch is set to "Yes," all text will be drawn as actual text. However, if the switch is set to "No," each text object in the sample program will be drawn as two parallel lines to indicate the text's location.

The advantage to showing only parallel lines is that the drawing can then be redrawn much faster. Also, if it is not necessary to view the text, the screen looks much less cluttered with parallel lines than with text.

HATCH

The Hatch option in the SWITCHES Menu allows the designer to change the show hatch lines switch. If the show hatch lines switch is set to "Yes," then whenever the screen is redrawn, all the currently defined hatch lines are shown. If the switch is set to "No," the hatch lines are not shown. This option is a convenient way to temporarily turn off the hatch lines, reduce screen clutter, and also speed up many of the operations of the program. It operates in a way similar to that of Words above.

The Hatch feature of the sample 2D program is discussed at length in Chapter 9.

DEBUG

The Debug option in the SWITCHES Menu allows the designer to turn the "debug" switch on or off. This switch is generally used only if the program appears to be working improperly. The information that is displayed when this option is set to "Yes" can help programmers to determine the source of an error.

ALL LEVELS

The All Levels option in the SWITCHES Menu allows the designer to specify whether or not to override the current levels configuration and unconditionally work with all levels. When the All Levels switch is set to "Yes," all levels are turned on and displayed, regardless of the levels configuration (see below). Setting the switch to "No" causes the program to use the established levels configuration. The initial value of the All Levels switch is "Yes."

LEVELS

The Levels option in the SWITCHES Menu allows the designer to specify which level or levels are to be turned on and which are to be turned off. When a particular level is set to "Y" (for yes, show that level) that level is visible and drawn on the screen. Conversely, when a level is set to "N" (for no) all objects in that level are "invisible" and not drawn on the screen, even though they still exist in the workfile. The initial value for all 250 levels available in the sample program is "Y."

When this option is selected, the program displays the Levels submenu as shown in Figure 4.8. The options in the Levels submenu enable the

Figure 4.8 The Levels submenu of the SWITCHES Menu.

designer to turn a particular level on or off, move forward or backward one level, and to move quickly through the 250 levels available—by jumping to a specific level, moving in either direction, moving 10 or 40 levels at a time. The direction, current level, and configuration of all of the levels are displayed at the bottom of the text screen.

5

ADDING OBJECTS

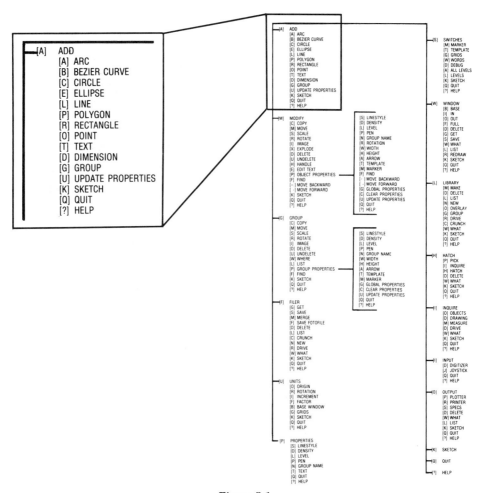

Figure 5.1

ADD

All 2D CAD systems provide a basic set of functions and operations
that enable the designer to create and then place objects—lines, circles,
polygons, text, and so on—on the graphics screen.

The ADD option in the sample program is used by the designer to
select objects to be created and then added to a drawing. Figure 5.2 shows
the ADD Menu. The ADD option allows the designer to create the elementary objects that compose a complex drawing. These elementary objects
include lines, rectangles, regular polygons, circles, arcs, ellipses, Bezier
(French) curves, dimensions, and text.

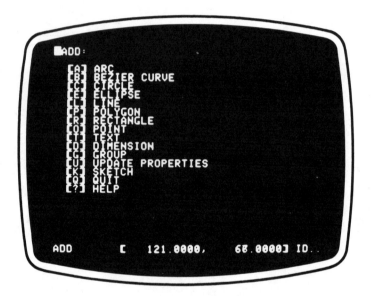

Figure 5.2 The ADD Menu.

The ADD Menu lists all of the primitive graphics objects that the
sample 2D program can create. The designer selects the object to be created
from the menu and it is interactively drawn on the screen as directed by the
movement of the cursor.

The digitizer is used to place an object's characteristic points. This is
done by moving the cursor to a specific location with the digitizer, and then
pressing the cursor button. The object is fixed or "accepted" when the
final characteristic point associated with that object is placed.

Hitting the ESCAPE key while in any ADD option immediately aborts
that option and returns the sample program to the ADD Menu. In addition,
if at any time during the creation of an object the designer wishes that
object to blink, he or she simply selects the object tracking function. This

causes the blinking to start. In some systems object tracking is known as "attentioning" an object.

When an object is being created in the ADD function of the sample 2D program, a submenu is displayed on the alternate text screen. This submenu indicates what special functions can be performed on an object that is presently being created. To select one of the special functions, the designer simply chooses the letter between the appropriate set of brackets.

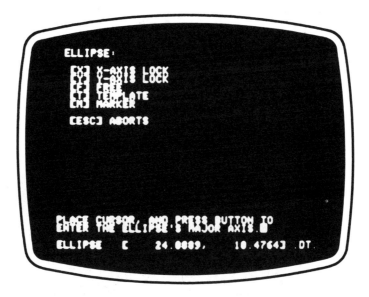

Figure 5.3 Special functions available for drawing an Ellipse.

The special functions that appear repeatedly for the various options in the ADD Menu include:

Arrowhead: causes an arrowhead to appear on one end of a line, arc, Bezier curve, or within a circle.

Template: causes the object to become a template object and to appear on the screen with a dotted rather than solid line formation.

Marker: causes a marker to be displayed at the center of an object.

X and Y Axis Lock: causes the object to become parallel with either the X or the Y axis, as indicated by the designer.

In each instance, hitting the identical key for a second time turns the special feature off. When unique special functions appear for a particular primitive object, they are described in detail as they occur.

ARC

The Arc option in the ADD Menu allows the designer to create a circular arc. Because of their characteristic shape, arcs can be defined in a variety of ways in different systems. In the sample 2D program arcs can be defined in two different ways. The first method defines an arc by entering three points on the arc: the first and second endpoint of the arc, and then a third arbitrary point on the arc. The second method defines an arc by entering the arc's center and two endpoints.

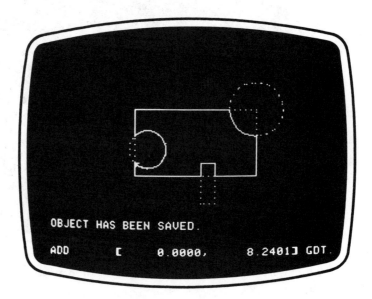

Figure 5.4 A drawing utilizing Arcs.

When the Arc option is initially selected by the designer, the first method of input automatically appears on the screen. If the second method is desired, the designer must select it before defining the first characteristic point.

Regardless of the option selected, the program does not begin to draw an arc until the first two characteristic points have been defined. These points are defined by moving the cursor to the desired locations on the screen and then pressing the cursor button. Once these two points have been defined, and object tracking or "attentioning" is enabled, the program repeatedly displays an arc. When the configuration of the arc is correct, the designer accepts it by pressing the cursor button.

The Arc function contains several unique special options:

Direction: causes the Arc to be drawn in the opposite direction. (This option is meaningful only when the second method for defining an arc is used.)

Free: allows the radius of the arc to be determined by the position of the cursor. (The program is in this mode when the three points arc option is first selected—and is available only when that option is selected.)

Radius: allows the precise radius to be specified—by entering it through the keyboard—when entering an arc by the three-points method.

BEZIER (FRENCH) CURVE

The Bezier option in the ADD Menu enables the designer to create one or more Bezier curves. In the sample 2D program a Bezier curve is defined by four characteristic points: the first and second endpoints of the curve, and the first and second curve control points (the points that determine the actual shape of the curve).

The sample program does not begin to draw a Bezier curve until the two endpoints have been defined. These points are defined by moving the cursor to the desired locations on the screen and then pressing the cursor

Figure 5.5 Special functions available for drawing a Bezier curve.

button. Once these two points have been defined, and object tracking has been enabled, the program repeatedly draws a Bezier curve whose shape is dependent on the location of the two control points.

Moving the cursor now moves both control points. The program allows the designer to move both control points, or either the first or the second control points, by selecting the appropriate special option. The program always draws from the first endpoint to the first control point, to the second control point, and then to the second endpoint. By moving the two control points separately, the designer can create interesting shapes. When the Bezier curve is satisfactory, the designer accepts it by pressing the cursor button.

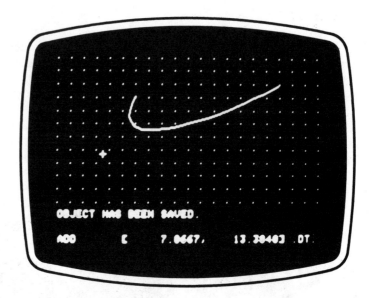

Figure 5.6   A Bezier curve.

Most systems provide special features for linking together several primitive objects in combination. Bezier curves can be linked together to appear as if they form one continuous curve with many inflection points. This can be done as follows:

1. Create a Bezier curve. Be sure that the markers are turned on. Draw a template line through the second control point and the second endpoint of the Bezier. It is best if the line extends all the way across the screen.

2. Create a second Bezier. Be sure that the beginning point of the second Bezier is the same as the ending point of the first Bezier. The second endpoint of the second Bezier can be placed wherever desired.

3. Position the first control point of the second Bezier curve in such a way that it lies on the dotted guide line.

4. Position the second control point of the second Bezier wherever desired.

5. The resulting Bezier curves should appear to be continuous.

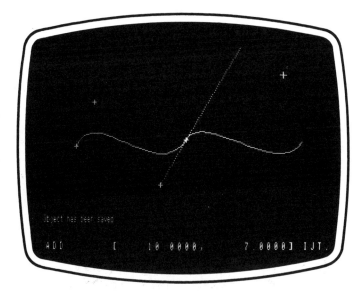

Figure 5.7 Two continuous Bezier curves.

In addition to the standard options, the Bezier function contains the following unique special options:

First: causes the first control point to follow the cursor and fixes the second control point at its current location.

Second: causes the second control point to follow the cursor and fixes the first control point at its current location.

Both: causes both control points to follow the cursor. The program is in this mode when the ADD Bezier option is first selected.

Place: fixes the location of both of the control points and allows the cursor to be moved around freely without changing the shape of the curve.

Line: draws a guideline that can be used to help create continuous Bezier curves.

CIRCLE

The Circle option in the ADD Menu enables the designer to create a circle. In the sample 2D program, circles can be defined in either of two different ways. The first method defines a circle by entering the circle's center and its radius. The two characteristic points in this method are the center point of the circle and the point identifying the location of the radius. The second method defines a circle by entering the diameter of the circle. The two characteristic points in this method are the two endpoints of the diameter of the circle.

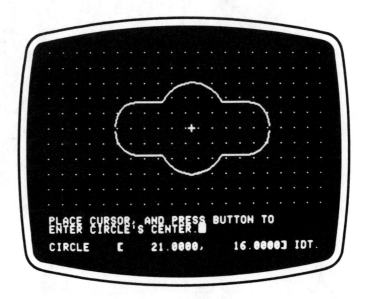

Figure 5.8 Drawing a Circle—entering the center point.

When the designer initially selects the Circle option in the ADD Menu, the first method of input described above is automatically displayed on the screen. If the second method is desired, the designer must select it before defining the first characteristic point of the circle.

Regardless of which method for defining a circle is selected, the program does not begin to draw a circle until the first characteristic point has been defined—either the center of the circle, or one endpoint of the diameter. The first point is defined by moving the cursor to the desired location on the screen and then pressing the cursor button. If object tracking is turned on, the program repeatedly draws a circle. Then move the cursor to identify the second point. When the circle size and location are correct, the designer accepts it by pressing the cursor button.

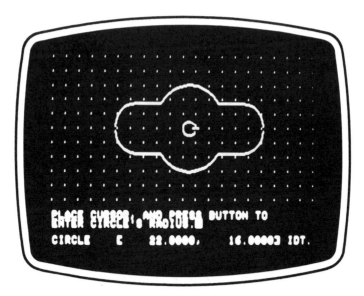

Figure 5.9 Drawing a Circle—entering the radius.

ELLIPSE

The Ellipse option in the ADD Menu enables the designer to create an ellipse. In the sample 2D program ellipses can be defined in either of two ways. The first method defines an ellipse by entering three characteristic

Figure 5.10 Drawing an Ellipse.

points on the ellipse: the two endpoints of the minor axis, and one endpoint of the major axis. The second method defines an ellipse by entering the ellipse's center and points on both the minor and major axis. The three characteristic points of this method are the center of the ellipse, one endpoint of the minor axis, and one endpoint of the major axis.

When the designer initially selects the Ellipse option in the ADD Menu, the first method of input described above appears on the screen. If the second method is desired, the designer must select it before defining the first characteristic point.

Regardless of which method is selected for defining an ellipse, the program does not begin to draw an ellipse until the first two characteristic points have been defined. These two points are defined by moving the cursor to the desired locations on the screen and then pressing the cursor button each time. Once these two points have each been defined, and object tracking has been turned on, the program repeatedly draws an ellipse. When the shape, rotation, and location of the ellipse are correct on the screen, the designer accepts it by pressing the cursor button.

LINE

The Line option in the ADD Menu enables the designer to create one or more lines. In the sample 2D program a line is defined by its two characteristic points: the first and second endpoints.

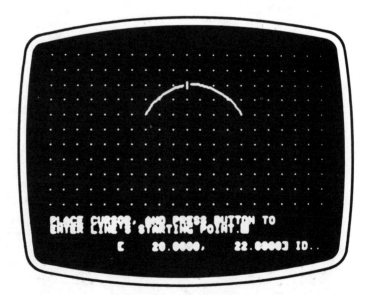

Figure 5.11 Drawing a Line—entering the starting point.

The program does not begin to draw a line until the first endpoint has been defined. The first endpoint is defined by moving the cursor to the desired location on the screen and then pressing the cursor button. Once the first endpoint has been defined, and object tracking has been turned on, the program repeatedly draws a "rubber-band" line from the first endpoint to the present cursor location. Moving the cursor changes the line's length and rotation. When the line's length, rotation, and location are correct, the designer accepts it by pressing the cursor button.

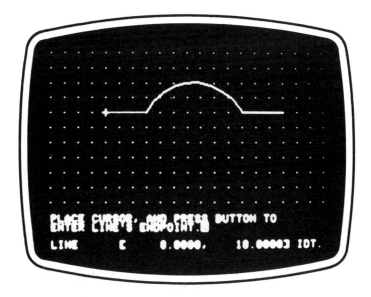

Figure 5.12   Drawing a Line—entering the endpoint.

A second line can then be drawn from the second endpoint of the previously created line to the now present cursor location. In the sample program this process of drawing lines continues until the designer terminates the Line option.

In addition to the standard options described above, the designer has several unique special options available for drawing lines.

Detach: allows the designer to redefine the first endpoint of the current line being drawn, thus providing a vehicle for drawing discontinuous (or detached) lines. Once the line is "detached," the designer continues on with normal line input.

Erase: enables the designer to remove the *last* accepted line. (Using the ESCAPE key causes *all* lines drawn during the current ADD Line option to be deleted.)

POLYGON

The Polygon option in the ADD Menu allows the designer to create a regular polygon. Although polygons can be created in a variety of ways in different 2D programs, in the sample 2D program a polygon is defined by two characteristic points. The first characteristic point defines the center of the circle within which the polygon is inscribed—and is essentially the center of the polygon. The second characteristic point defines the radius of the circle within which the polygon is inscribed—and is essentially the radius of the polygon.

When the designer selects the Polygon option in the ADD Menu, the program requires that the number of sides of the polygon be determined first. In the sample program a polygon can have from 3 to 30 sides.

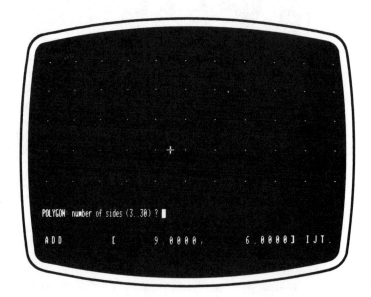

Figure 5.13 Drawing a Polygon—determining the number of sides.

The program does not begin to draw a polygon until the first characteristic point, the center of the polygon, has been defined. The center is defined by moving the cursor to the desired location on the screen and then pressing the cursor button. Once the center has been entered, and object tracking has been turned on, the program repeatedly draws a polygon.

Moving the cursor about the center of the polygon changes the radius of the polygon. Moving the cursor away from the center causes the polygon to grow; moving it toward the center causes the polygon to shrink. The polygon can also be rotated by moving the cursor around the center. When the size, rotation, and location of the polygon on the screen are correct, the designer accepts it by pressing the cursor button.

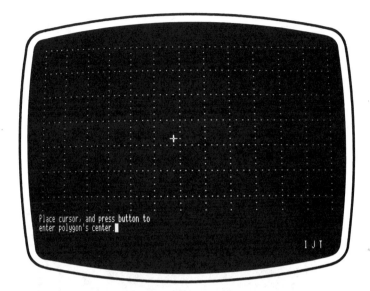

Figure 5.14 Drawing a Polygon—entering the center point.

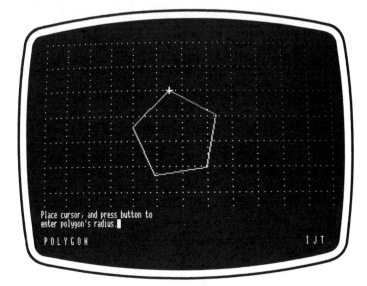

Figure 5.15 Drawing a Polygon—entering the radius.

RECTANGLE

The Rectangle option in the ADD Menu enables the designer to create a rectangle. In the sample 2D program a rectangle is defined by two characteristic points: one corner and then the opposite corner of the rectangle.

The program does not begin to draw a rectangle until the first corner has been defined. The first corner is defined by moving the cursor to the desired location on the screen and then pressing the cursor button. Once the first corner has been defined, and object tracking has been turned on, the program repeatedly draws a rectangle. The first corner is now fixed and the designer must then determine the opposite corner. Moving the cursor changes the location of the opposite corner. When the size and location of the rectangle are correct, the designer accepts it by pressing the cursor button.

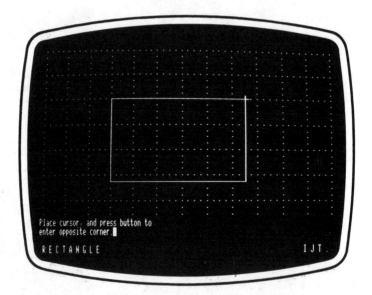

Figure 5.16 Drawing a Rectangle.

POINT

The Point option in the ADD Menu allows the designer to add one or more points to a drawing. To add a point the designer identifies a desired location and presses the cursor button to accept it. This action is then repeated for each point to be entered.

A point is always plotted on the plotter as a single dot. Points assigned a template property, however, do not plot at all—although they do appear on the screen as template points.

TEXT

The Text option in the ADD Menu enables the designer to create and place multiple lines of text on a drawing. In the sample 2D program, text is defined by a single characteristic point: the starting point of the text, the lower left-hand corner.

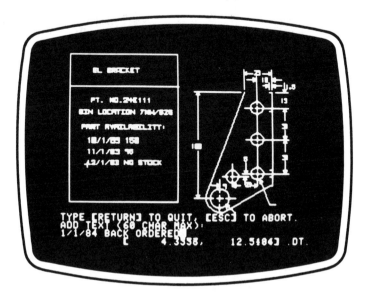

Figure 5.17 Entering Text.

Text can be entered and placed in a variety of ways in the different 2D programs available. In the sample program, when the Text option in the ADD Menu is selected, the program prompts the designer to enter a line of text up to 60 characters in length. The text is typed in through the keyboard, to a maximum of 60 characters. When the line of text is complete, the designer accepts it by hitting RETURN (ENTER, on some other systems), and the program then actually draws the text.

If any mistakes are made while typing in the text, it can be edited first. Hitting the backspace (or left arrow) key deletes the last character entered. Hitting ESCAPE aborts the current ADD Text option. It is important to note that when the ESCAPE key is pressed, the program removes *all* lines of text added during the current ADD Text operation.

Once the line of text has been entered, the designer positions it by moving the cursor to the desired location on the screen and then presses the cursor button to accept it. After entering its position, the program again prompts the designer to enter another line of text from the keyboard.

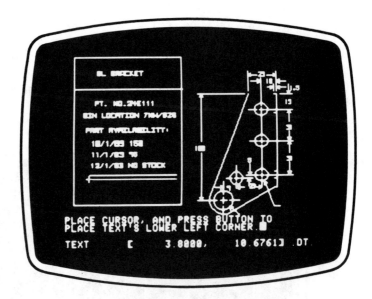

Figure 5.18 Using the cursor to place Text on the screen. (Text is displayed as two parallel lines until accepted.)

Hitting *only* the RETURN key in response to a prompt to enter another line of text terminates the ADD Text operation.

Many 2D CAD systems provide the designer with a variety of text font options. The sample 2D program does not; it generates a single text font for all characters.

The ADD Text option does include two special functions that facilitate the editing and placement of text.

Rotate: causes the text to be rotated 90 degrees counterclockwise. Selecting this option a second time moves the text to a rotation of 180 degrees; a third time, to 270 degrees; a fourth time, returns the text to 0 degrees.

Edit: allows the designer to edit text just entered.

DIMENSION

The Dimension option in the ADD Menu allows the designer to dimension any object on the screen. Dimensions can be entered and defined in a variety of ways in different 2D systems. In the sample 2D program a dimension line is defined by three characteristic points. The first and second points define the first and second endpoints on the object being dimensioned. The third point defines the height of the dimension, or in other words, it defines the length of the leader lines of the dimension.

The sample program does not begin to draw a dimension line until the first two characteristic points have been defined. These two points are

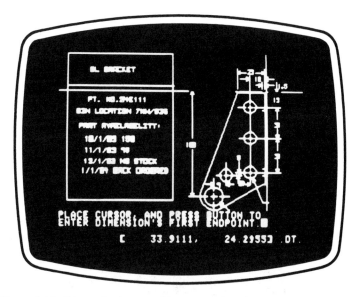

Figure 5.19 Dimensioning an object—entering the dimension endpoint.

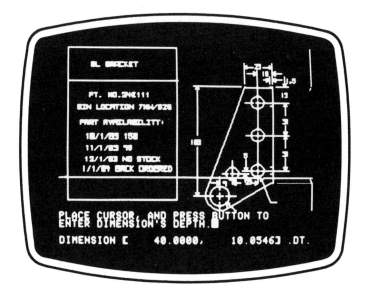

Figure 5.20 Dimensioning an object—entering the dimension depth.

defined by moving the cursor to the desired locations on the screen and then pressing the cursor button. Once these two points have been defined, and object tracking has been turned on, the program repeatedly draws dimension lines.

Moving the cursor away from the object being dimensioned lengthens the leader lines; moving the cursor closer shortens them. The lines can be flipped from one side of the object to the other by moving the cursor to the other side of the object being dimensioned. The two endpoints just entered are fixed and cannot be changed. When the size and location of the dimension lines are correct, accept them by pressing the cursor button.

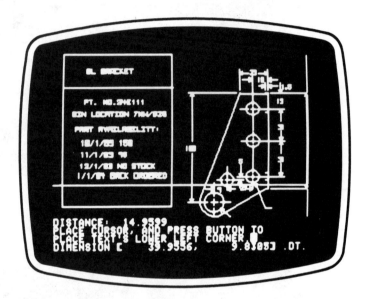

Figure 5.21   Dimensioning an object—placing the dimension text.

As soon as the dimension is accepted, the program generates a text displaying the numeric value of the dimensioned distance. Moving the cursor moves this text. When the text is placed correctly, the designer accepts it by pressing the cursor button.

Characteristics of dimensioning vary from system to system. In the sample 2D system the dimension leader lines do not touch the object being dimensioned; when they are plotted they are 1/16 of an inch away from the object. Also, the arrowheads on the dimension bar are plotted at a length of 5/32 inch and have an included angle of 15 degrees. Other systems may provide a variety of options for creating leader lines and for specifically determining the style of the arrowheads.

A variety of special options are available to assist the designer in dimensioning.

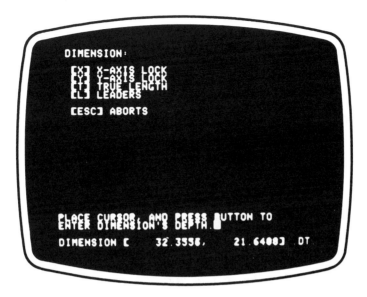

Figure 5.22 Special functions available for dimensioning.

X Axis: measures the horizontal distance of the object.

Y Axis: measures the vertical distance of the object.

True Length: measures the true length of the object.

Leaders: allows the designer to selectively turn off one or both of the leader lines.

A variety of additional options are available to assist the designer in creating and placing dimension text.

X Axis: causes the dimension text to be oriented horizontally along the X axis.

Y Axis: causes the dimension text to be oriented vertically along the Y axis.

Rotate: causes the dimension text to rotate in degrees as defined in the Rotation option of the UNITS Menu.

True: causes the dimension text to maintain the same angle as the dimension lines.

Break: breaks the dimension line at the center to allow dimension text to be inserted.

Outside: places the dimension lines outside the area being dimensioned; that is, the arrowheads of the dimension are pointing inward instead of outward.

Decimal: causes the dimension text to be represented as a decimal number. For example, 68.5930 would be a typical decimal number. Before producing the decimal dimension, the program asks the designer to determine the number of significant decimal digits—up to 4. Entering a "0" would cause the decimal number above to become 68. Entering a "2" causes the dimension to become 68.59.

Feet and Inches: causes the dimension to be represented as feet and inches. For example, the same 68.5930 would be represented as 68 feet.

GROUP

Many 2D CAD systems feature the ability to create and utilize standard symbols (or "groups" used as standard symbols), either created by the designer or purchased with the software. The Group option in the ADD Menu allows the designer to add one or more complex groups or symbols to a drawing.

Symbols are added to drawings by selecting them out of previously created "symbol libraries." Symbol libraries are usually displayed on a portion of the designer's digitizer pad so that they are readily available for inclusion in a drawing. In the sample 2D program a group or symbol is defined by two characteristic points. The first point entered selects the symbol to add; the second point entered defines the placement of the symbol.

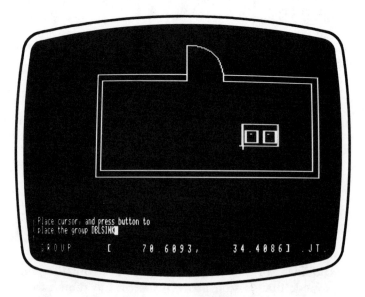

Figure 5.23 Adding a Group (or Symbol) to a drawing.

A symbol can be selected in either of two ways. If the program input mode is set to "digitizer," the symbol is selected by placing the cursor button over the desired pictorial representation on the symbol overlay and then pressing the cursor button; the same operation can be achieved with a stylus. If the program input is set to "keyboard," the symbol is simply selected by typing in its location number in the symbol library.

When object tracking is turned on, the program repeatedly displays that symbol. Moving the cursor moves the symbol, changing its location on the screen. When the location is correct, the designer accepts it by pressing the cursor button. Additional symbols can be added to a drawing by simply repeating this process.

Several unique special options are available to the designer to aid in the placement of symbols in a drawing.

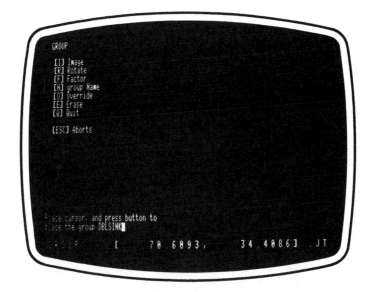

Figure 5.24 The Group (or Symbol) submenu in the ADD Menu.

Rotate: causes the symbol to rotate 90 degrees. Each time this option is selected the symbol rotates an additional 90 degrees.

Image: causes the symbol to be imaged about the Y axis.

Define: allows the designer to define a new, temporary handle point by which to drag the symbol.

Handle: changes the handle point of the symbol to the point defined in the library as the handle point.

Scale: allows the designer to visually scale the symbol.

Factor: allows the designer to enter a numerical scaling factor.

Override: overrides all property values stored with the symbol and causes the symbol to use the currently defined properties. The program rounds the dimension to the nearest 1/16 inch.

Inches: causes the dimension to be represented as inches. For example, the same 68.5930 would be represented as 5 feet $8^{9}/_{16}$ inches. The program rounds the dimension to the nearest 1/16 inch.

Edit: allows the designer to override what the program produces as the dimension text. The program prompts the designer to enter the desired dimension—up to 60 characters followed by RETURN.

UPDATE PROPERTIES

When objects are created initially, their characteristic properties are defined or established by the designer. In the course of developing a design it may become necessary or desirable to alter one or more of the established properties for an object to be created. Most systems provide a variety of ways for updating the primary properties or characteristics of an object or a group of objects.

The Update Properties option in the sample 2D program enables the designer to change the global properties of any object or group and to adjust the smallest increment. When this option is selected by the designer, the program displays the current global properties. A typical display is shown in Figure 5.25. Once these properties are displayed, the designer is

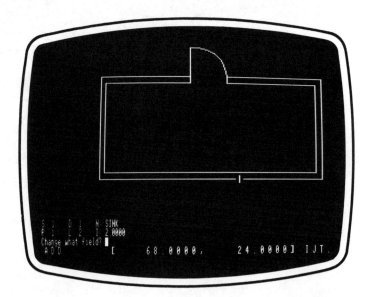

Figure 5.25 The Update properties option in the ADD Menu.

expected to select the global property to be updated—Linestyle, Density, Level, Pen, Color, Increment, Group Name. Changing the properties using the Update Properties option also changes the properties found on the PROPERTIES Menu—so Update Properties serves to offer a quicker method of inquiring into and changing one or more global properties.

GUIDELINES

Some systems feature an option that places guidelines on the screen. Guidelines assist in the precise placement of objects on the screen. In the sample 2D program guidelines can be created by using the long crosshairs or by using template lines.

FILLET

Some 2D programs have features that enable the designer to create one or more fillets or rounds between lines, circles, and arcs. Fillets and rounds are created in a variety of ways in the different 2D programs. In some programs a fillet is created by first identifying the two objects bounding the fillet and by then defining the radius of the fillet. Figures 5.26 and 5.27 illustrate filleted objects. The designer must first identify the first, and then the second object bounding the fillet. When both objects have been defined, the program produces a fillet between the two objects.

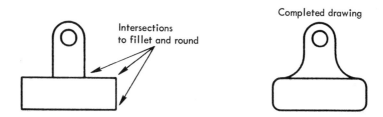

Figure 5.26 Example 1: Using the Fillet option in the ADD Menu.

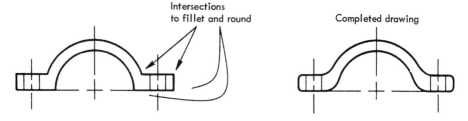

Figure 5.27 Example 2: Using the Fillet option in the ADD Menu.

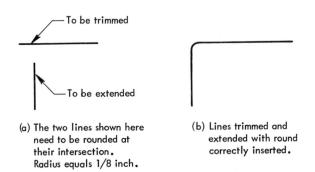

(a) The two lines shown here
    need to be rounded at
    their intersection.
    Radius equals 1/8 inch.

(b) Lines trimmed and
    extended with round
    correctly inserted.

Figure 5.28 Example 3: Using the Fillet option in the ADD Menu.

Programs may feature either manual or automatic trim. Automatic trim causes the program to automatically trim the objects bounding the fillet once the fillet has been created. Objects are trimmed to flow tangent into the fillet. Manual or no trim allows the designer to touch the portion of the object to be trimmed back to the fillet once the fillet has been created. Two lines that are trimmed and rounded are illustrated in Figure 5.28.

6

2D INPUT, OUTPUT, AND FILE MANAGEMENT

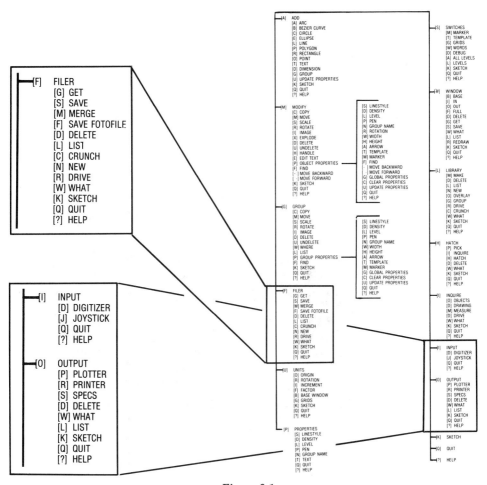

Figure 6.1

INPUT

There are now a variety of common input devices used in 2D CAD systems, in addition to the computer keyboard itself, including a digitizer pad (also known as a graphics tablet), joystick, stylus, roller ball, thumb wheel, and so on. These input devices enable the designer to send signals to the computer that are interpreted as either actions to be performed by the 2D CAD program, or actions that lead to the creation and manipulation of graphic symbols or representations on the screen. The most common input devices are digitizer pads and joysticks. Less common are input using a plotter and various input techniques using the keyboard. These are not features of the sample program, but are nonetheless presented at the end of the INPUT section.

The OUTPUT function enables the designer to produce a hard copy of a current drawing. The hard copy can be sent to a plotter—or in some systems, either a plotter or a graphics printer. A variety of plots can be created, depending on the type of plotter being used in the system. By creating tailor-made sets of plotter specifications, the designer can even determine exactly what part of a drawing is to be placed where on the plotter and to what scale.

In the sample 2D program, the INPUT Menu appears as shown in Figure 6.2. Selection of an option from this INPUT Menu changes the current device used for graphics input.

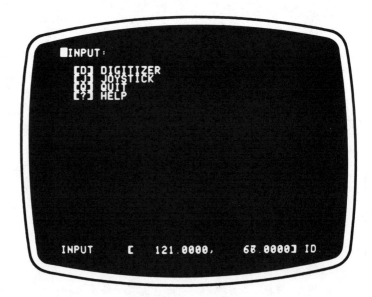

Figure 6.2 The INPUT Menu.

DIGITIZER

The Digitizer option in the INPUT Menu tells the program that the designer is using a digitizer pad (or graphics tablet) for graphics input. When the designer selects this option, the program displays a confirming message.

JOYSTICK

A second popular input device, popular primarily for its exceptional low cost, is a joystick. Using a joystick is economical, but its accuracy is not as good as that of the digitizer pad.

To overcome the deficiencies of the joystick the program provides a special feature, the joystick scaler. When the joystick scaler is selected, the movement of the cursor is restricted to one-fourth of the screen area presently accessible. This key can be pressed again and again until the cursor has been placed on the desired screen dot.

KEYBOARD

In some 2D CAD programs there is an input option—sometimes referred to as "Keyboard"—that tells the program that the designer will be using the keyboard arrows or the thumb wheel to do graphics input. When such an option

Figure 6.3 A microcomputer with a thumbwheel.

is selected, the program displays a confirming message and the cursor moves to the center of the screen. Pressing any of the four arrow keys causes the cursor to move in the direction specified by that arrow key.

A thumb wheel can be used in place of the arrow keys. Turning the wheel clockwise is equivalent to hitting the right arrow key—it moves the cursor to the right. A counterclockwise rotation is equivalent to hitting the left arrow key—it moves the cursor to the left.

PLOTTER

Some 2D CAD programs provide an option that tells the program that the designer will be using a plotter to do the graphics input. Selecting this option, when available, allows the designer to put a drawing on the plotter, and then to use an optical sighting device to position the pen holder to act as a digitizer.

BOUNDARIES

When designers use a digitizer pad for input, it is common practice to actually "digitize" (or trace) a drawing or also to use one or more symbol libraries on the digitizer pad. When the designer wishes to digitize a drawing, or when symbol libraries are used in this way, the designer must somehow tell the 2D program what the actual boundaries of the active surface of the digitizer pad are to be.

A Boundaries, or similar, option is used in conjunction with the digitizer pad and permits the designer to specify the "active" digitizing area on the digitizer tablet.

OUTPUT

In the sample program the OUTPUT Menu appears as shown in Figure 6.4. The options in the OUTPUT Menu perform a variety of functions that enable the designer to produce a hard copy of a drawing.

PLOTTER

The Plotter option in the OUTPUT Menu enables the designer to draw any portion of the current drawing onto a graphics plotter. Exactly how the drawing is drawn on the plotter is determined by a set of "plotter specs." There are two types of "plotter specs," the default specifications and the

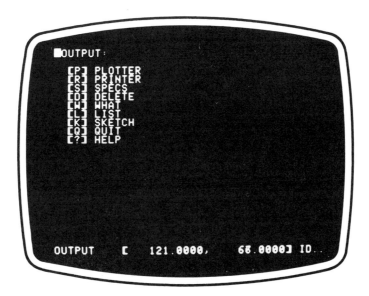

Figure 6.4 The OUTPUT Menu.

user-defined specifications. The default plotter specs maps the current screen window—that is, the program calculates a default scaling factor appropriate to the size of the plotter being used—onto the total surface area of the plotter.

The user-defined plotter specs map the current drawing onto the plotter in any way that the user defines. Although more complicated than the default specs, the user-defined plotter specs are much more powerful. The user-defined plotter specs are created using the Specs option of the OUTPUT Menu (see below).

Plotter capabilities vary considerably from one plotter to the next, so it is important that the designer become fully familiar with the capabilities and limits of the plotter used in his or her system.

PRINTER

Many 2D programs also provide the designer with an option to print the current screen image on a dot matrix printer. A printer option provides the designer with an opportunity to obtain inexpensive, relatively high-quality output from a graphics printer. Most programs require a specific hardware configuration to use the printer option successfully.

Since this is a low-resolution form of graphics output, it is best used to get a rough hard copy to proof a drawing. Also, in most cases, the program prints exactly what is on the screen without interpretation; that is, program commands, guidelines, templates, and so on, as well as the actual drawings are printed when they appear on the screen.

SPECS

As noted above, the Specs option in the OUTPUT Menu enables the designer to define a set of customized "plotter specs." Plotter specs are used to determine exactly *what* part of a drawing is placed *where* on the plotter, and to *what scale*.

The plotter specs can be thought of as defining a mapping function that determines how the objects in the real-world coordinate system of a drawing are to be placed on the limited coordinate system of the plotting surface. The mapping formula used is:

(real-world coordinate system) $\times$ (scaling factor) =

(plotter coordinate limits)

In such a formula, the "real-world coordinate limits" represent the screen window, and the "plotter coordinate limits" represent the plot limits. By assigning values to any two of the three factors in the mapping function, the third factor can be computed automatically. This is exactly what is done when defining plotter specs.

Figure 6.5 shows the Specs Menu for the sample 2D program. When the Specs option is selected, the computer displays a standard set of plotter

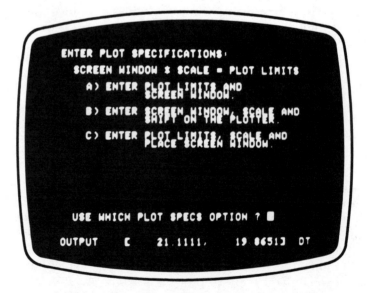

Figure 6.5 The Plotter Specs submenu in the OUTPUT Menu.

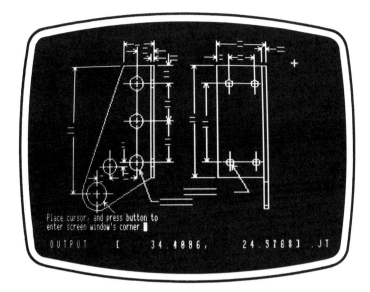

Figure 6.6 Using Plotter Specs—the drawing window.

specs on the screen. The three components of a plotter specs are the screen window, the plot boundaries, and the scaling factor. Both the plot boundaries and the scaling factor are displayed as numeric quantities. The screen window is shown as a rectangle on the screen.

Once the plotter specs are on the screen, the designer can change any of the components of those specs. Usually, the changing of one component requires an equivalent "adjustment" in one of the other components. When this is the case, the program displays the additional component option or options that must then be adjusted. For example, changing the scale factor that is applied at plot time requires that either the plotter boundaries or the screen window be recalculated. Once the designer changes the scaling factor, the program then asks which of the two additional components is to be adjusted, the boundaries or the screen window.

When the Specs option is selected from the OUTPUT Menu, the program displays, together with the Specs Menu, a variety of additional features and options.

A Metric option enables the designer to change the input screen to reflect a metric unit of measure.

A Window option allows the designer to change the screen window component of a plotter specs. When this option is selected, the program begins blinking the current screen window and displays the following options on its own submenu: Move, Scale, Place, Unproportional. The Move option allows the designer to move the current screen window. The Scale option allows the designer to scale the current blinking window. The Place option allows the designer to move the cursor without moving the screen window. The Unproportional option is used in conjunction with Scale option and allows the program to scale unproportionately the blinking screen window.

Remember that if the size or scale of the screen window was changed, the program requires the designer to indicate whether the boundaries or factor should be recalculated.

DELETE

The Delete option in the OUTPUT Menu allows the designer to delete any plotter specs defined previously. When this option is selected, the program requires the designer to enter the name of the plotter specs to be deleted. When the name is entered, the program next requires the designer to reaffirm the delete command before proceeding to delete the plotter specs from the workfile.

WHAT

As the number of saved plotter specs increases, it becomes likely that the designer will not be able to remember the precise characteristics of each set. The What option in the OUTPUT Menu of the sample program enables the designer to review all the information contained in a selected set of plotter specs. When the selected plot specs are found, the program displays three pieces of information:

1. The screen window that is to be mapped onto the plotting surface will be shown as a blinking rectangle.
2. The limits of the plotting surface onto which the screen window will be mapped.
3. The scaling factor that will be applied when the screen window is mapped onto the plotting surface.

LIST

The List option in the OUTPUT Menu allows the designer to list all of the plotter specs that have been saved in the workfile. List operates here in a fashion similar to the List option in the FILER Menu. List permits the designer to view the names of saved plotter specs so that the appropriate plotter spec can be retrieved for use in a drawing.

FILER: 2D FILE MANAGEMENT

All drawings created by the designer on any 2D general-purpose drafting
system—or any CAD system, 2D or 3D—must be permanently saved on a
hard disk or diskette for filing purposes and be available for other types of
manipulation. The diskette on which 2D drawings are saved is called a
drawing file disk. In the sample 2D program, the FILER—or comparable—
Menu provides a Save option that can transfer the drawing that is in the
workfile to a desired drawing file disk. It also provides a Get—or retrieve—
option that can retrieve a drawing from a drawing file disk and store it in
the workfile. Composite drawings can be created by retrieving more than one
drawing from the designer's drawing file disks by successively using the Get
option.

In the sample 2D program FILER options can be used at any time
during the creation of a drawing. Figure 6.7 shows the FILER Menu.

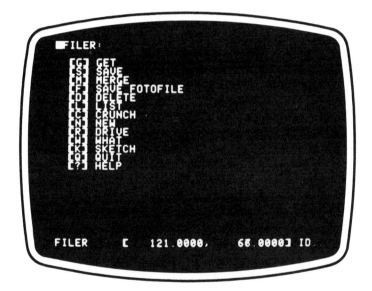

Figure 6.7 The FILER Menu.

GET

All 2D programs provide the designer with a simple means of saving drawings
and then retrieving them from diskette. The Get option in the FILER Menu
in the sample 2D program allows the designer to transfer a previously saved
drawing from a drawing file diskette to the current workfile. In the process

of retrieving a drawing, the program clears the workfile of all objects before copying the requested drawing into the workfile.

When this option is selected, the program prompts the designer to identify the name of the program to be retrieved. The program then searches the drawing data file. When the named drawing is found, the program displays a confirming message.

SAVE

The Save option in the FILER Menu allows the designer to save the current workfile onto the drawing diskette. When this option is selected, the program responds with a prompt that requires the designer to name the drawing to be saved. Once the name has been entered, the program displays a message confirming that the drawing has been saved. If, for any reason, the workfile cannot be saved to the diskette, the program responds with an appropriate message.

MERGE

CAD contributes to increased productivity by enabling the designer to build and develop complex drawings from previously saved drawings. The Merge option in the FILER Menu allows the designer to merge a previously saved drawing from the drawing file disk drive with the current information in the workfile. When this option is selected, the program prompts the designer to enter the name of the drawing file to merge. The designer must next determine whether to use the Units and Properties of the current drawing in the workfile or to preserve those of the drawing file to be merged. When the decision is made and entered, and the drawing has been completely read in from the disk, the program displays a confirming message.

SAVE FOTOFILE

Some of the more sophisticated 2D programs have a feature that enables the designer to save an exact copy of the image presently on the graphics screen —sometimes called a "screen dump." The fotofile is a special type of file that can be used by other programs, but not by the 2D drafting program. There is no compatability with the 2D program because the fotofiles have no information associated with them except a dot pattern.

Basically, then, a fotofile is nothing but a picture of the screen at a particular instant in time. These photos can be used in a cycle presentation much like a slide show. For example, an animation sequence can be produced using the 2D program to draw the picture frames, and then save them as fotofiles.

DELETE

The Delete option in the FILER Menu of the sample program allows the designer to delete any 2D drawing saved onto diskette. When this option is selected, the designer enters the name of the drawing to be deleted. The program then displays a message confirming that the drawing file has been deleted. If the program is unable to find the drawing file indicated, an appropriate message is displayed.

LIST

Although diskette storage space is limited, it is still advisable for the designer to establish some logical system for coding and naming drawings. Most 2D CAD systems provide the designer with a means to identify saved drawings by listing the names of those drawings on the graphics screen.

In the sample 2D program, the List option in the FILER Menu allows the designer to list all of the 2D graphics files that have been saved on a diskette. Additional information about the drawing file provided by the

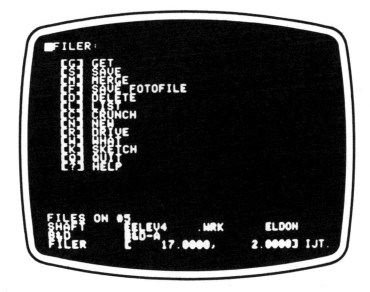

Figure 6.8 The FILER Menu with a listing of saved drawings displayed at the bottom of the screen.

program includes the date it was created, the total number of objects in the file, the total number of different symbols in the drawing, and the average number of objects for each different symbol. These last three fields can be very helpful when a drawing is being transferred into the workfile. They provide information that can assist the designer in judging whether the existing file size is adequate to accommodate the new drawing file. Similar options in other systems provide comparable information about a drawing.

The List option itself is quite helpful when the designer cannot remember the name of a particular graphics file that is to be deleted or retrieved.

CRUNCH

The process of adding and deleting objects from a drawing can leave empty —and unusable—"holes" in the workfile. The Crunch option in the FILER Menu, or comparable options in other programs, should be used only after careful thought. After the workfile has been crunched, the possibility of "undeleting" any objects, text, or groups that have previously been deleted is gone. In most programs, when this option is selected, a warning message is first displayed on the screen to prevent the designer from accidentally "crunching" the workfile.

NEW

The New option in the FILER Menu allows the designer to initialize a workfile. When New is selected by the designer, the program erases everything from the workfile and clears the graphics screen. All objects, saved windows, and plotter specs are erased. Generally, this option is used by the designer whenever he or she is ready to start a new drawing. If the drawing presently in the workfile is of any continued value, it should always be saved by the designer before starting a new workfile. Once the New option is completed, nothing can be recovered through the Undelete option. When the New option is selected, a warning message is displayed on the screen and the designer must confirm the program command to continue.

DRIVE

The Drive option in the FILER Menu enables the designer to reconfigure the system and to override the default data drive—the drive where the work diskette resides—established by the program. This function is used in systems where a third or fourth disk drive has been added, or in systems employing a Winchester-type high-density disk drive.

WHAT

The workfile is of fixed size. The What option in the FILER Menu reports the available workfile storage and the amount of workfile storage used by the current drawing. The storage units are related to object "records." A record can be thought of simply as an object or a text item (some long text items require two records of storage).

As drawings become more and more complex it becomes increasingly important for the designer to be aware of how close the critical features of the system are being approached. By their very nature microcomputer-based CAD systems are limited in their storage capacity. The What option —and similar options in other CAD programs—displays useful information about the status of the current workfile.

When the What option is selected, the program displays the maximum amount of space available in the workfile for objects and the default data drive number. The What option does not, however, give any information about how full a particular diskette may be.

This program option eventually tells the designer when the workfile is full. If the workfile does become full, the designer can use the Crunch option of the FILER Menu to create additional room.

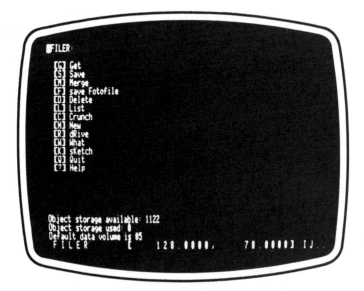

Figure 6.9 The What option in the FILER Menu.

ARCHIVE

Many systems that utilize high-density hard disk drives and accommodate the creation and storage of larger and more complex drawings allow the designer to maintain "archived" drawings. In particular, they allow the designer to store and to retrieve large drawings stored on one disk to and from several smaller disks.

On systems with a hard disk, some means of "backing up" drawings from the hard disk to a floppy disk is essential. This also means there must be some provision for "backing up" a drawing stored on the hard disk that will not completely fit on one floppy disk. An Archive or similar option performs this hard disk-to-floppy disk backup.

BACKUP

Many systems also contain an option that enables the designer to make a backup copy of an entire workfile. This option is especially useful when the program is running with a workfile on a memory volume instead of on a disk volume. In this case, a backup option protects against any unexpected machine failures that might destroy the memory volume.

FILE MANAGEMENT GUIDELINES

The foregoing sections describing features and functions of a typical FILER Menu relate primarily to the technical aspects of file management for the sample 2D system that has been chosen as the focus of this text. In addition to the technical aspects, a CAD operator must be concerned about the management of the information that is stored in the files on the disks or diskettes.

Considerable problems can occur if proper management practices are not exercised. Good "housekeeping" practices include frequent backup of workfiles so that valuable work is not lost if a hardware or operator malfunction should destroy the information in a file. Another good practice is never to use a master file as a workfile, whether the file is a program or a drawing file.

Generally, hard disks require more sophisticated file procedures since the information is always on the hard disk available for access, while the diskettes are normally removed from a computer while not in use. Appropriate procedures include copying off those portions of the disk which have been changed during a drawing session to tape or to a diskette. Some installations have software programs designed specifically to handle the task of copying or backing up the hard disk. These programs may copy all of the files on the hard disk or may automatically detect the files that have been changed and copy only the changed files.

These collections of drawing files are sometimes referred to as a data base. Other subjects related to the management of drawing files include administration of the data base, security of the data base, and various methods of retrieval and reporting from the data base. The interested reader should consult a book on data base administration for a more advanced treatment of this topic.

7

MODIFYING A 2D DRAWING

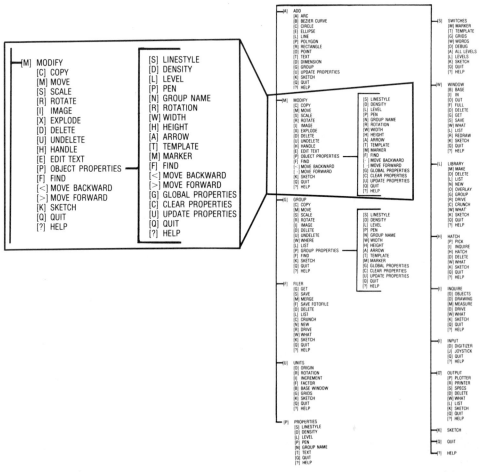

Figure 7.1

MODIFY

Although a designer may take great care in working with a 2D program to create a drawing, often a drawing may not look exactly right. A circle may need to be moved, a line may need to be rotated, or a rectangle may need to be copied and scaled down. In all these cases, options in the MODIFY (or comparable) Menu provide the means to produce the desired results. The MODIFY function in a 2D program allows the designer to graphically manipulate any of the objects currently displayed on the screen.

In the sample 2D program the MODIFY option can be reached from the MAIN Menu by selecting the "M" character. The MODIFY Menu appears as shown in Figure 7.2. "Handles" (or handle points) are used frequently in the MODIFY options of the various 2D programs. The handle point is that point on an object that is used by the program to move, scale, or rotate that object.

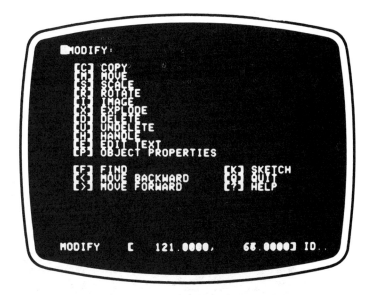

Figure 7.2 The MODIFY Menu.

2D CAD programs maintain a "list" of all the objects in a drawing. When objects are created, they are added to the end of the list. When the designer enters the MODIFY Menu from the MAIN Menu, the program "points to" the last object in the list, assuming that the designer wants to modify the last object added.

The program must point to an object for that object to be modified. To determine which object is being pointed to, the designer must turn on the tracking (attentioning, in some systems) mode and the program causes the object to blink repeatedly. The blinking continues until the designer turns object tracking off, or until the designer causes another object to be pointed to by the program.

COPY

The ability to copy an object, once or repeatedly, and to place that copy in a drawing contributes directly to the productivity gain of CAD. The Copy option allows the designer to make one or more copies of an object. When this option is selected, the object pointed to by the program is duplicated. If the tracking mode is turned on, the blinking copy then follows the cursor across the screen.

In the sample 2D program the copy is always attached to the cursor by its handle point. The designer positions the cursor at the desired location, and accepts the copied object by pressing the cursor button. Once accepted, another copy is automatically produced, and it, too, now follows the cursor across the screen. Again, the designer positions the cursor and accepts the copied object. The Copy option continues to produce multiple copies until the designer terminates it by quitting the Copy option. All of the copies produced are stored in the workfile. Hitting the ESCAPE key aborts the Copy operation and erases *all* of the copies made.

The options available to the designer to copy and locate an object on

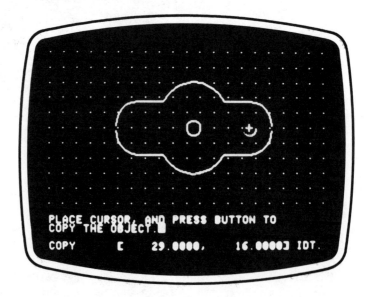

Figure 7.3 Using the Copy option in the MODIFY Menu.

the screen—and those options generally available for the options in the MODIFY Menu—are:

Handle: changes the handle point from the primary to the secondary handle point for those objects that have two handle points.

X Axis Lock: causes the cursor to read only the X-axis change.

Y Axis Lock: causes the cursor to read only the Y-axis change.

Free: frees the designer from the X or Y mode and causes the object to follow the movements of the cursor. The program is in this mode when the Copy option is first selected.

Original: causes the object to revert to its original location.

Place: fixes the location of the object and allows the cursor to be moved without the object following it.

A number of special options are available to perform functions unique to only certain options within the MODIFY Menu. These special options are described as they occur. In the Copy option there are two such special options:

Repeat: allows the designer to repeatedly copy the object in one direction, two directions, or circularly about a selected point. Depending on the option selected, the designer is required to enter the total number of copies (for one direction), or the total copies in the X and

Figure 7.4 The Copy Repeat submenu of the Copy option.

Figure 7.5 Using Copy Repeat—determining the number of copies to be created.

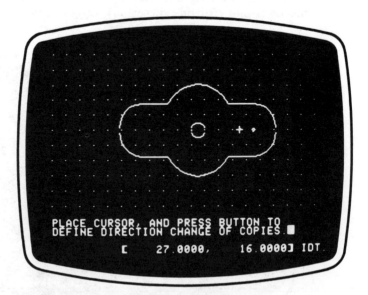

Figure 7.6 Using Copy Repeat—defining the direction change for the copies.

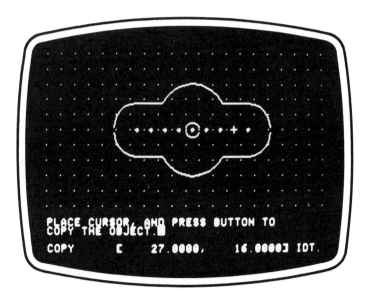

Figure 7.7 Using Copy Repeat—entering the copies.

in the Y direction (for two directions). For the circular option the designer is required to provide the relative angular offset (in degrees) between each of the copies, the total number of copies, and the relative angular offset to be applied to each copy—usually this entry is the same as the first (for circular). The program then draws a specified number of copies at a specified rotation from a defined center point.

Erase: causes the last accepted copy of the object to be erased.

MOVE

One of the common features of all 2D CAD systems enables the designer to modify a drawing by selecting an object—or series of objects—to be relocated or moved from one position to another.

The Move option in the MODIFY Menu of the sample program allows the designer to move an object across the screen. When the Move option is selected, the handle point of the object pointed to by the program moves with the cursor, dragging the object with it. When the new location of the object is correct, the designer accepts it by pressing the cursor button. It now remains fixed in its new location. Hitting the ESCAPE key returns the object to its original location.

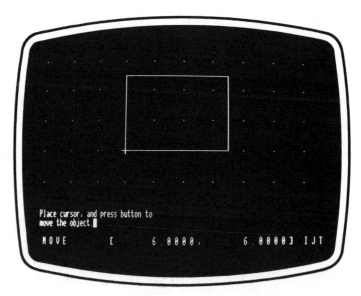

Figure 7.8 Using the Move option in the MODIFY Menu.

SCALE

Most 2D CAD programs provide the designer with features or functions to identify particular objects in a drawing and to change the size, or scale, of those objects.

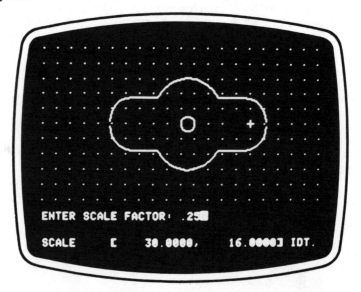

Figure 7.9 Using the Scale option—entering a scaling factor.

The Scale option in the MODIFY Menu of the sample program allows the designer to change the scale (or size) of an object. When the Scale option is selected, the same scaling factor is applied in both the X and Y directions—so the object is not distorted when scaled. The designer scales each object by fixing a stationary point and then scaling it by pulling outward (to make the object larger) or inward (to make the object smaller) from a handle point on the object. Consequently, moving the cursor scales the object relative to its handle point.

Hitting the ESCAPE key returns the object to its original size.

In addition to the special options common to most of the options in the MODIFY Menu, the Scale function contains an additional special option designed to facilitate scaling an object:

Factor: enables the designer to enter a numeric scaling factor from the keyboard. A factor greater than 1.0 increases the size of the object. A factor smaller than 1.0 decreases the size of the object.

ROTATE

Most 2D programs contain a feature or function that enables the designer to select an object and to change its configuration in relationship to the X or the Y axis.

The Rotate option in the MODIFY Menu of the sample program allows the designer to rotate an object. When this option is selected, the object

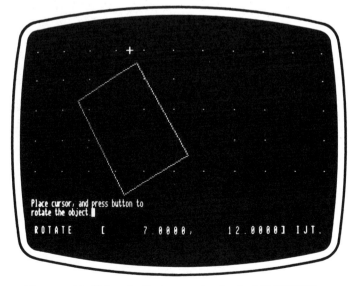

Figure 7.10 Using the Rotate option in the MODIFY Menu.

pointed to by the program rotates about its handle point as guided by the movements of the cursor. The object's size remains unchanged, and its handle point remains fixed. When the rotation of the object is correct, the designer accepts it by pressing the cursor button. Hitting ESCAPE returns the object to its original position.

IMAGE

CAD systems increase productivity by having built-in features that handle repetitive tasks with great efficiency. Objects that are symmetrical can be partially drawn and then completed by using a process called imaging.

The Image option in the MODIFY Menu of the sample program enables the designer to create a mirror image of an object about a given line. When the Image option is selected, the program displays a blinking horizontal line on the screen. This is the image line, the line about which the object is imaged. This line can be vertical, horizontal, or rotated at any angle and can be moved anywhere on the screen. When the image line is positioned correctly, the designer accepts it by pressing the cursor button. The program then draws the imaged object.

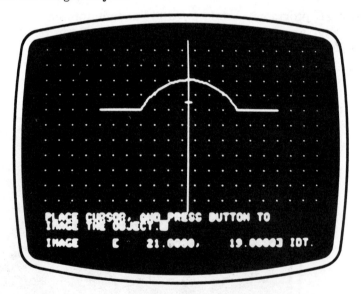

Figure 7.11 Using the Image option in the MODIFY Menu.

The special options available to the designer to facilitate imaging an object are:

X Axis Lock: causes the image line to become horizontal. The program is in this mode when the Image option is first selected.

Y Axis Lock: causes the image line to become vertical.

Rotate: causes the image line to rotate about a fixed point. The position of the cursor just prior to selecting the Rotate option becomes the point about which the image line is rotated. Once this option is selected, moving the cursor no longer moves the image line, but causes it to rotate about the fixed point instead.

Move: causes the image line to follow the cursor across the screen. The angle of the image line remains fixed. If the image line were being rotated, selecting Move would fix the angle of the line, and moving the cursor would no longer cause the line to rotate. The program is in this mode when the Image option is first selected.

Copy: allows the designer to switch between imaging the object and imaging a copy of the object. This is a powerful feature of the system that quickly enables the designer to "compile" a drawing with minimum input.

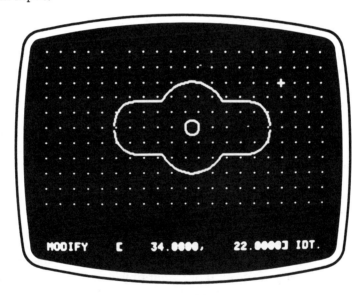

Figure 7.12 The imaged object.

EXPLODE

A few 2D systems feature an option that enables the designer to select an object—for example, a circle—and to separate or explode that object into even more primary objects—for example, a circle becomes a series of continuous lines.

The Explode option in the MODIFY Menu enables the designer to explode an object into a set of line segments. When the Explode option is selected, the program "explodes" the object being pointed to into a set of continuous line segments. These line segments are then added to the workfile as individual lines. Thus, once an object has been exploded, each of the individual line segments produced can then be modified.

Any object in the drawing can be exploded. When Explode is selected by the designer, the program prepares to modify the first line segment of the exploded object. In the sample program, even a line can be exploded. A line is broken at its midpoint into two identical lines, each one-half the size of the original. This allows the designer to trim a line in order to insert other objects.

When Explode is selected, the original object remains in the workfile but is no longer associated with the drawing. The Undelete option (see below) can be used to return this object to the drawing.

DELETE AND UNDELETE

All 2D programs provide the designer with a function that enables him or her to selectively eliminate objects from a drawing. A few programs even provide a function that enables the designer to recall and replace in the drawing previously deleted objects.

The Delete option in the MODIFY Menu of the sample program allows the designer to delete an object from the drawing. When the Delete option is selected, the program deletes the object being pointed to from the screen. The object remains in the workfile but is no longer associated with the drawing.

The Undelete option in the MODIFY Menu of the sample program allows the designer to return this object to the drawing. When the Undelete option is selected, the program looks for any objects that have previously been deleted. If it does not find any, it immediately leaves the Undelete option and returns to the MODIFY Menu. However, if it does find some undeleted object(s), it repeatedly draws that object and responds with a prompt asking the designer whether or not the object should be "undeleted." Answering "Yes" causes the object to be undeleted. Answering "No" causes the program to skip over that object and to search for another deleted object. The program continues to search for and point to previously deleted objects until none are left in the workfile or until the designer terminates the Undelete option.

MOVE BACKWARD AND MOVE FORWARD

In order to modify an object—change its size or location or rotation—the designer must first indicate to the program which object is to be modified. Programs provide a variety of features for designers to locate objects to be modified in a drawing.

The Move Backward and Move Forward options in the MODIFY Menu of the sample program are also provided for the purpose of locating an object to be modified. When they are created, objects are stored by the program in an imaginary file or list. The list is "circular" and if either the move forward or move backward arrow key is hit a sufficient number of times, the end of the list is reached and the list is "recycled." These arrow keys tell the program what object is to be modified, so when the correct object is reached, the cycle is stopped. In effect, they enable the designer to step through the list of objects in a drawing.

These two keys provide a means of sequentially stepping through the objects in the workfile until the one to be modified is found. Each time that the Move Backward or Move Forward arrow key is pressed, the program moves one object forward or backward in the list.

If object tracking is on, objects blink on the screen. When the Move Forward key is selected, the object currently being pointed to stops blinking and then another one, the next one in the list, starts to blink. The object now blinking was "forward" in the list of objects. The Move Backward key operates in the same fashion. If object tracking is not turned on, the object being pointed to by the program blinks only once.

FIND

In the sample 2D program the Find option allows the designer to quickly search for and find an object to be modified. When the Find option is selected, the designer moves the cursor to any point on the object to be modified and then presses the cursor button. The program then searches through the drawing looking for the touched object. It can take a few seconds for the program to find this object, especially if the drawing has many objects; the more complex the drawing the longer it may take. When the program has completed its search, it blinks the object—if object tracking (or attentioning) is turned on.

The program allows the designer to be about three screen dots away from the object and still find it. If the wrong object starts blinking, the designer was not close enough to the target object. When placing the cursor on an object, the designer must try not to place it at a location close to another object. The program has an easier time finding the correct object when the location is isolated.

EDIT TEXT

All 2D CAD systems provide the designer with a means to enter text and with one or more ways to change or edit text previously entered. The Edit Text options in the MODIFY Menu allow the designer to perform various modifications to any text object.

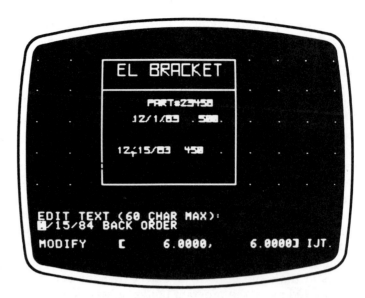

Figure 7.13 Using the Edit Text option in the MODIFY Menu.

The Edit option within this Edit Text Menu allows the designer to edit text on the graphics screen. The text on the screen must first be pointed to by the program—in the same fashion that objects must be pointed to before they can be modified.

When the Edit option is selected, the program displays the actual line of text toward the bottom of the screen. The designer enters the new text from the keyboard. Typing the backspace (left arrow) key erases one character at a time from the end of the text.

Hitting the ESCAPE key enables the designer to abort the Edit operation and causes the program to redisplay the original text.

Hitting RETURN accepts the new text.

Some of the more sophisticated 2D programs provide a number of additional options for editing text, including justify, scale, and adjust.

The Justify option in the Edit Text submenu allows the designer to justify any text object on the screen. To justify text means to align the text with a predefined "justification line." For example, if the designer is drawing a graph and wants all the labels along the vertical axis to start in the same row, the justify option should be used.

Text can be right, left, center, and bottom justified.

Right: The right-hand edge of the text is lined up with the justification line.

Left: The left-hand edge of the text is lined up with the justification line.

Center: The center of the bottom edge of text is lined up with the justification line.

Bottom: The bottom edge of the text is lined up with the justification line.

The Scale option allows the designer to scale any text object on the screen. The scaling factors that are applied in the X and Y directions may, in this instance, be different, allowing the designer to change the proportion of text height and width.

When this option is selected, the program begins redrawing the currently blinking text to its new scale. Moving the cursor pulls one end of the text and thereby changes its scale. When the new scale is correct, the designer accepts it by pressing the cursor button.

The Adjust option in the Edit Text submenu allows the designer to adjust any upside-down text object so that it can be read.

OBJECT PROPERTIES

Regardless of the CAD system in use, each object that a designer creates has certain primary properties—attributes or characteristics that define an object. All primitive objects have several properties. For example, two properties of a line are a group name of "N1" and a plotter linestyle of 3.

Each 2D system has some means of enabling the designer to modify or change the primary properties of an object once it has been created. In the sample program the designer uses the Object Properties option in the MODIFY Menu to selectively change one or more of the properties of any object.

The object that is "eligible" for changing is the one currently being pointed to by the program. If object tracking is turned on, this object blinks repeatedly. The designer can move through the list of objects and change the object being pointed to by the program by using the Find, Move Backward, or Move Forward commands.

The basic idea behind the Object Properties option is that an object's properties are changed by the designer creating a list of proposed changes, and then changing the current values of the object to the proposed values all at once. Once a list of proposed changes has been created, the designer can modify the properties of any number of objects by typing a single key for each object.

In the sample program the Object Properties Menu appears as shown in Figure 7.14. When the Object Properties option is selected for a particular object, the program lists all of that object's properties on the left side of the screen. The list is broken down into two columns: a "CURRENT" column which contains the current values of all' the object's properties,

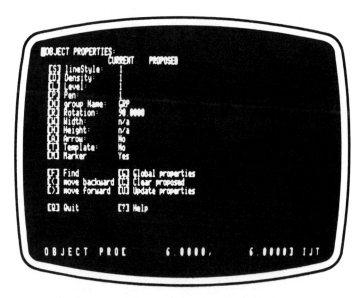

Figure 7.14 The Object Properties submenu in the MODIFY Menu.

and a "PROPOSED" column which contains the proposed new values for those properties. Initially, the proposed column is blank.

When the designer selects an option from the Object Properties Menu, the program positions the designer to enter the new proposed values of that property. After it is entered, that new value is displayed in the PROPOSED column. Any one or several new values can be entered in the PROPOSED column.

It is important for the designer to remember that the proposed values are just that, proposed. In the sample program, the current values are not changed to the proposed values until the Update Properties option within the Object Properties Menu is selected. Properties with no proposed changes are not changed. Other programs, of course, may provide different methods for identifying and then changing an object's properties.

The Find, Move Backward, and Move Forward options in the Object Properties submenu operate in the same way as they do in the MODIFY Menu.

Linestyle

Many programs support more than one screen and plotter linestyle—solid, dotted, dashed lines, and so on. The Linestyle option in the Object Properties submenu enables the designer to change the screen and/or plotter pen linestyle property of an object.

Density

One way of varying line width in a 2D system is to cause the plotter to strike or draw over a line two or more times. Each time the line becomes a bit thicker, as well as somewhat darker in comparison to the other lines.

The Density option in the Object Properties submenu enables the designer to change the number of times that the plotter overstrikes each line in an object.

Varied line thickness can also be achieved by using pens of varied thickness in a multipen plotter.

Level

Most 2D programs allow the designer to work with several levels (or layers) at one time. The number of levels available to the designer may vary from system to system, but the principle of using levels is always the same—like objects or objects with common properties or functions should always be placed on the same level.

The Level option in the Object Properties submenu allows the designer to change the level property of an object.

Pen

Many plotters allow the designer to use multiple pens to accomplish color coding or varied line thickness. The Pen option in the Object Properties submenu enables the designer to change the plotter pen property of any object.

Group Name

In most 2D systems common objects, or objects that create a more complex object or symbol, are grouped together under a single "group name." Groups can then be modified or manipulated in the same way that objects can (see the section GROUP in Chapter 8). The Group Name option in the Object Properties submenu enables the designer to change the group name property of an object.

Rotation

As a designer works through a complex drawing it sometimes becomes apparent that a primary or text object is not in the correct relationship with either the X or the Y axis.

The Rotation option allows the designer to change the rotation of an object. Since the exact desired rotation is entered in degrees, this option is an extremely powerful one for the precise placement of objects on the screen.

Width and Height

Text widths and heights are usually established from the outset by setting certain default parameters before a complex drawing is begun. The character text width defines how wide a single character is; height, of course, defines how high. The width and height are always in terms of real-world coordinates.

For example, if a drawing ranges from 0 to 10 feet in the X direction, and 0 to 8 feet in the Y direction, a character text width and height of 1 causes the text to be 1 foot high and wide.

The text Width and Height options in the sample 2D program enable the designer to make adjustments in text width and height while in the midst of a drawing.

Arrow

Arrowheads can be placed on an object—arcs, lines, Bezier curves—when they are created or can be placed on them later through the Object Properties (or comparable) menu in a 2D program. The Arrow option allows the designer to either add or remove an arrowhead from an object.

Template

Any object can be given template status when it is first created, or later through the use of the Template option in the Object Properties (or comparable) Menu in a 2D program. The Template option allows the designer to determine whether or not an object is drawn in template status.

Marker

The Marker option allows the designer to change whether or not an object's center marker is to be shown.

Global Properties

As noted at the outset of this section, Object Properties are changed by setting up a proposed column of changes. The Global Properties option enables the designer to set up this PROPOSED column containing all the current global property values. Once this column is established, the designer proceeds to determine which properties need to be changed.

Clear Proposed

The Clear Proposed option does precisely what the name implies; it clears the PROPOSED column of any changes. This option is used by the designer if a mistake is made in the PROPOSED column.

Update Properties

It is not uncommon in a 2D system for the program to have embedded commands that actually cause changes to take place. In the sample 2D program, changes in object properties do not actually take place until the designer selects the Update Properties option. The Update Properties option tells the program that the established changes have been reviewed by the designer and are correct. Since it might be necessary to change several objects to the new object properties, the Update Properties option also enables the designer to scroll the list of objects. Each time an object is blinked on the screen that needs to be modified, the designer simply hits the "U" for Update Properties, thereby telling the program that this object needs to have its object properties changed to those newly established ones.

Color

Some microcomputer-based CAD systems support multiple colors that can be displayed on the screen.

A Color option—or similar option—allows the designer to change the screen color of an object. When this option is selected, the program requires the designer to enter the new screen color number. Of course, this option is meaningful only for those systems that support color screens.

8

2D GROUPS AND WINDOWS

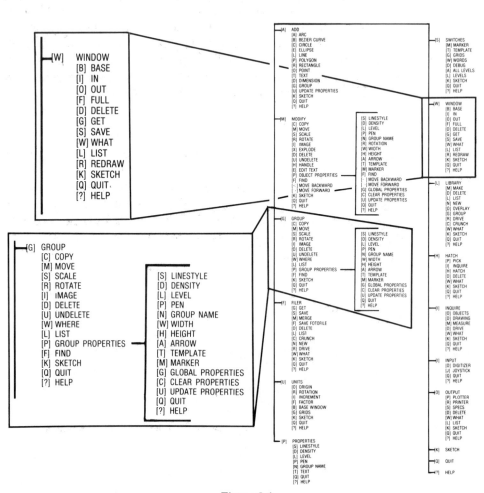

Figure 8.1

GROUP

A "GROUP" is an arbitrary, user-defined collection of objects. Groups provide a convenient way of manipulating the basic components of a complex drawing. Groups are common to all CAD systems. Groups allow the designer to move, scale, copy, and otherwise manipulate large numbers of symbols at one time. Groups also provide a useful means of storing these complex figures on diskette, where they can be used as components of future drawings. The options available in a GROUP or comparable menu enable the designer to perform these manipulative tasks.

A group is identified by an arbitrary name determined by the designer. When a drawing is first started, all objects that are created are assigned a global default group name by the program. In any 2D program there is a mechanism for changing the default name to a name more meaningful to the designer—in the sample 2D program it is done using the options in the PROPERTIES Menu.

Objects within a drawing may form logical groupings. For example, a small portion of an overall drawing may consist of several objects grouped to form a piping or electrical network. At some time, it may be necessary to move, copy, scale, or otherwise modify this network as a unit. By giving all of the objects within the network the same group name, the designer can work with this collection as a group.

In the sample program the GROUP Menu appears as shown in Figure 8.2. Several of these options—Delete, Undelete, Sketch, Quit—have already

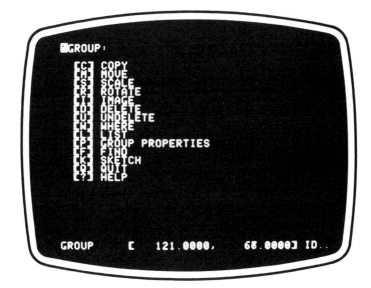

Figure 8.2 The GROUP Menu.

been discussed under the ADD or MODIFY Menu. These options are performed in essentially the same manner for groups as they were for objects; it is simply groups of objects that are being acted upon rather than single objects. Copy, Move, Scale, Rotate, and Image are sufficiently complex to merit additional discussion below.

Most options in the GROUP Menu require the entry of a group name. This is necessary so that the program knows which group is to be modified. When certain options in the GROUP Menu are selected (e.g., Copy, Move, Scale, Rotate, Image), the program prompts the designer to provide the group name. When a group name is entered, the program attempts to verify that the group actually exists in the workfile. If it does exist, the program continues on with the selected option. If it does not exist, the program tells the designer so and asks for a new name to be entered.

In certain instances (e.g., when modifying a group) a program may display a submenu of options along the side or at the bottom of the screen. These submenus give the designer greater control over the selected GROUP option. In the sample program these "Options" are displayed at the top of the text screen.

Figure 8.3 Displaying the Group Name.

COPY

Any group can be developed into a complex symbol that can be used over and over again in the same drawing. The Copy option in the GROUP Menu gives the designer the ability to make one or more copies of an existing group and to place these copies anywhere on the graphics screen. For each copy requested, the designer has the option of providing a new group name.

Copies of a group are always moved by the handle point. The program allows the designer to determine the location of the handle point for each group. The handle point facilitates the precise placement of a group's copy or copies on the screen.

MOVE

The Move option in the GROUP Menu enables the designer to move an existing group from its present location to a new one anywhere on the screen.

Groups are moved from the handle point, and the handle point is determined by the designer when the Move option is first selected. The designer positions the cursor over the handle point desired and accepts it by pressing the cursor button. The designer then moves the cursor to the new location of the handle point and accepts this location by pressing the cursor button. The program then redraws the object at its new location.

SCALE

Groups are scaled in much the same way as primary objects. The same scaling factor is applied in both the X and Y directions, so the group is not distorted when scaled. In the sample program the designer first identifies a fixed stationary point and then scales the group by pulling the handle point.

When the Scale option in the GROUP Menu is selected, the designer must first identify the stationary point. The stationary point is entered by moving the cursor to the appropriate location and pressing the cursor button. Once the stationary point has been identified, the designer is required to determine the handle point.

The handle is the point that the program holds onto while the group is scaled. The designer enters the handle by moving the cursor to the desired location and then presses the cursor button. Once the handle point has been established, the program generates a blinking line between the handle and stationary points. This line represents the magnitude of the scaling and is called the scaling line. Moving the cursor causes the scaling line to grow or shrink in size. These first two points that were entered also serve to determine the rotation of the scaling line and that rotation cannot be changed. When the scale is correct, the designer accepts the group by pressing the cursor button and the program draws the group to its new scale.

The Scale option in the GROUP Menu also provides a keyboard option for entering a numeric scaling factor. This option operates in precisely the same way as that described for modifying objects.

ROTATE

The Rotate option in the GROUP Menu enables a designer to rotate a group, and can be used in conjunction with Copy, Move, and Scale. When the designer has entered the name of the group to be rotated, the program prompts the designer to determine a "pivotal point." The pivotal point becomes the point about which the group is rotated. The designer enters the pivotal point by moving the cursor to the desired location and then pressing the cursor button. The program next prompts the designer to enter the handle point.

The handle is that point which the cursor holds onto while the group is rotated. The designer enters the handle point by moving the cursor to the desired location and then presses the cursor button. Once the handle point has been determined, a blinking line appears between it and the pivotal point. The rotation is always relative to the two points that have been established. The designer moves the cursor about the rotation line and when it is correct, presses the cursor button and the program draws the group in its new rotation.

IMAGE

The Image option in the GROUP Menu is an extremely powerful operation that enables the designer to create mirror images of existing groups. When the designer has entered the name of the group to be imaged, the program

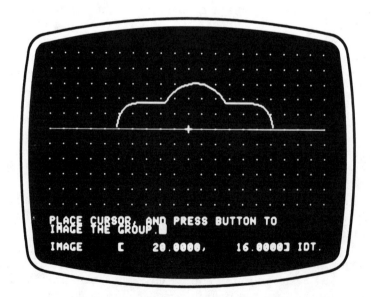

Figure 8.4 Using the Image option in the GROUP Menu.

displays a blinking horizontal line on the screen. This is an image line—the line about which the group is imaged. This line can be vertical, horizontal, or rotated at any angle, and can be moved anywhere on the screen. When the image line is positioned correctly, the designer accepts it by pressing the cursor button. The program then draws the imaged group.

When imaging, it is possible for the designer to instruct the program to image a copy of the specified group rather than the group itself. This option then produces two groups, the original and the imaged copy.

A variety of keyboard options are available for imaging a GROUP— X and Y Axis Lock, Rotate, Move, and Copy. The options operate in the same manner as they do when imaging an object.

WHERE

The more complex a drawing becomes, the more difficult it is to keep exact track of the location of a given group. The Where option in the GROUP Menu enables the designer to locate groups on the screen.

When this option is selected, the program requires that the designer enter the name of the group to be located. Once the appropriate name is typed in, the program searches through the workfile for that group. When the program finds the group, it begins blinking each object in the group. The designer hits RETURN to stop the blinking and to return to the GROUP Menu.

LIST

The List option in the GROUP Menu is a similar aid to the designer in working with a complex drawing. This option allows the designer to produce a list on the screen of the groups in a workfile.

GROUP PROPERTIES

All objects created in a 2D general drafting system have certain specific characteristics or attributes. Similarly, objects that form a group share a set of characteristics or attributes.

By using the Group Properties option the designer can selectively change one or more characteristics or attributes (i.e., the properties) of *all* the objects in a group. The basic idea behind the Group Properties option is that the designer changes a group's properties by creating a list of proposed changes. The current properties of all the objects in the group are then changed to the proposed properties all at once.

Like many of the options on the GROUP Menu, the Group Properties option requires the entry of a group name. This is necessary in order for the program to know which group's properties are to be modified. Once the name of the group is typed in by the designer, the program responds by printing a list of all the properties available for change. Unlike the Object Properties option, which displays current properties on the screen when it is selected, in Group Properties the list contains only one column—PROPOSED—which contains the proposed new properties when you enter them. Figure 8.5 shows the PROPOSED column initially—before any changes are entered.

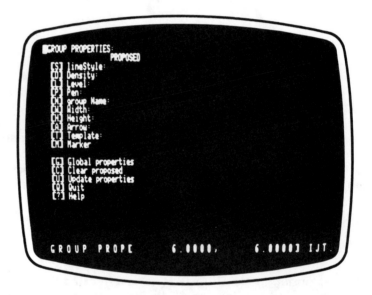

Figure 8.5 The Group Properties submenu in the GROUP Menu.

When an option in the Group Properties submenu is selected, the program positions the designer to enter the new proposed value of that property. The new value is then displayed in the PROPOSED column. It is important to remember that the proposed values are just that, proposed. The current values of the objects in the group are not changed to the proposed values until the Update Properties option is selected. When the Update Properties option is selected, all objects in the group have their properties changed to the PROPOSED values. Properties that have no changes proposed remain unchanged.

A brief description of each option in the Group Properties submenu follows.

Linestyle

The Linestyle option in the Group Properties submenu allows the designer to change the screen and plotter linestyle of all the objects in a group. The designer enters the new linestyle value and the program displays it in the PROPOSED column.

Density

The Density option in the Group Properties submenu enables the designer to change the number of times the plotter overstrikes each object in a group. The designer enters the new density value and the program displays it in the PROPOSED column.

Level

The Level option in the Group Properties submenu allows the designer to change the level property of all of the objects in a group. The designer enters the new level value and the program displays it in the PROPOSED column.

Pen

The Pen option in the Group Properties submenu allows the designer to change the plotter pen property of all the objects in a group. The designer enters the new pen value and the program displays it in the PROPOSED column.

Group Name

The Group Name option in the Group Properties submenu allows the designer to change the name of all of the objects in a group. The designer enters the new group name value and the program displays it in the PROPOSED column.

Width and Height

The Width and Height options in the Group Properties submenu enable the designer to change the character width and height of all of the text objects in a group. The designer enters the new value of either or both text width and height and the value(s) is displayed in the PROPOSED column.

Arrow

The Arrow option in the Group Properties submenu allows the designer to add or remove the arrowheads from all of the objects in a group. The designer enters the new value for arrowheads—either a "Y" for yes, or an "N" for no—and the program displays the new value in the PROPOSED column.

Template

The Template option in the Group Properties submenu allows the designer to make all of the objects in a group into template objects or to change those that are template objects into "normal" objects. The designer enters the new template value—either "Y" for yes or "N" for no—and the program displays the new value in the PROPOSED column.

Marker

The Marker option in the Group Properties submenu allows the designer to determine whether or not center markers are shown.

Global Properties

The Global Properties option in the Group Properties submenu allows the designer to set up a PROPOSED column containing all the global property values. When this option is selected, the program displays all of the global property values in the PROPOSED column.

Clear Proposed

The Clear Proposed option in the Group Properties submenu clears the PROPOSED column to contain no changes.

Update Properties

The Update Properties option in the Group Properties submenu actually sets and makes permanent all the changes listed in the PROPOSED column.

Color

Some programs feature a variety of screen colors as one of the characteristics or properties of both objects and groups. A Color—or similar—option allows the designer to change the screen color of *all* objects in a group.

WINDOW

The WINDOW Menu option, sometimes called "zoom" in other programs, allows the designer to manipulate the current viewing "window." To understand a window, imagine that the screen is a moving window through which the drawing is viewed. Windows are used to work on a portion of a drawing at a level of greater or lesser detail. They can be used to "blow up" or "shrink" the drawing that is displayed on the screen.

The options in the WINDOW Menu allow the designer to move, shrink, and magnify this viewing window. Figure 8.6 shows the WINDOW Menu. Windows are created by setting up a "window rectangle" and then moving that rectangle to surround the area of the screen to be examined. Both the creation and movement of the window rectangle are done using the cursor. Once the window is correctly positioned and the appropriate size, it is accepted and the windowed view of the drawing is then displayed by the program. To prevent distortion of the drawing, all window rectangles are created with the same proportions as the viewing screen.

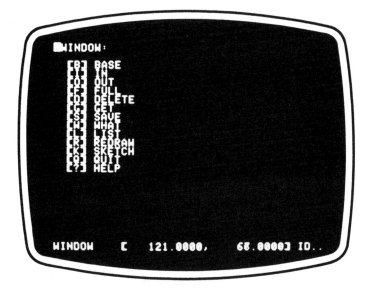

Figure 8.6 The WINDOW Menu.

BASE

In UNITS and PROPERTIES the designer establishes a set of screen coordinates that govern the use of the program for a particular drawing. The Base window is a reflection of that set of user-defined coordinates—it is defined by the Base option of the UNITS Menu. In the sample program the initial base window is 0-279 in the X direction and 0-209 in the Y direction.

When the designer first enters the 2D program, it is at the base window. It is the starting point at which all drawings are begun. During the course of a drawing the designer may have windowed "in" or windowed "out" many times, but the designer can return to the initial base window at any time from the WINDOW Menu.

IN

All 2D programs provide the designer with a feature that enables him or her to focus on a smaller portion of a drawing and to magnify that portion and display it on the entire screen. In some systems this is achieved by using a "zoom" feature. In other programs this feature is called windowing in. Windowing (or zooming) in gives the appearance of rapidly moving into a room—objects that appeared small and lacking in detail suddenly appear considerably larger and well defined.

When the Window In option in the WINDOW Menu is selected, the program requires the designer to enter the first corner of the window desired. The designer moves the cursor to the position of the first corner and then presses the cursor button. The program then repeatedly draws a rectangle on the screen. This rectangle is a representation of the window. The rectangle grows or shrinks as the cursor is moved. When the rectangle surrounds the portion of the drawing that is to be magnified, the designer accepts it by pressing the cursor button. The screen is immediately updated to show the magnified portion of the drawing.

The program does not allow the designer to Window In indefinitely. If the window requested would cause the program to lose numeric precision, the Window In request is denied.

Special keyboard options are available to allow the designer interactively to change and control the Window In function. These keyboard options are displayed on a menu line along the side of the screen. They are:

Scale: causes the blinking window to grow or shrink, depending on the cursor location. The program is initially in this mode when the Window In option is first selected.

Move: allows the designer to move the blinking window across the screen using the cursor. Moving the window does not change its size.

Place: fixes the size and location of the blinking window and allows the designer to move the cursor without affecting the window.

ESCAPE: aborts the Window In operation.

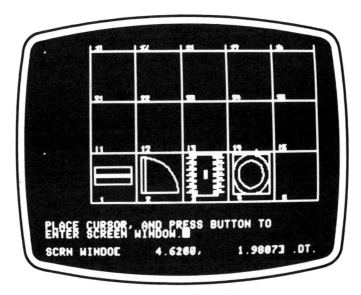

Figure 8.7 A portion of a Symbol Library.

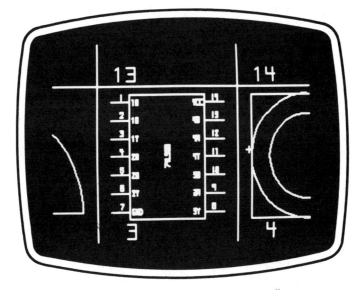

Figure 8.8 A Window In on Symbol #3.

OUT

All 2D programs provide the designer with a similar feature that enables him
or her to take the image displayed on the screen and shrink it into a selected
area of the screen. In some systems this is also achieved by using a "zoom"

feature. In other programs this feature is called windowing out. Windowing (or zooming) out gives the appearance of rapidly moving out of a room. Objects that appeared large and well defined suddenly appear considerably smaller and lose their definition.

When the Window Out option in the WINDOW Menu is selected, the program requires the designer to enter the first corner of the desired window. The designer moves the cursor to the position of the first corner and then presses the cursor button. The program then repeatedly draws a rectangle on the screen. This rectangle is a representation of the window. The rectangle grows or shrinks as the cursor is moved. When the rectangle surrounds the portion of the drawing that the designer wishes to shrink the screen into, the designer accepts it by pressing the cursor button. The screen is immediately redrawn. The original screen has now been reduced to fit into the area defined by the new window. Objects that were previously off the screen may now be visible.

The program does not allow the designer to Window Out indefinitely. If the requested window would cause the program to lose numeric precision, the Window Out request is denied.

The same special keyboard options that are available to allow the designer to interactively change and control the Window In function are also available to assist with the Window Out function.

FULL

When working on a complex drawing many components may be outside the coordinates of the present screen. The Full option in the WINDOW Menu creates and displays the smallest window that contains the entire drawing. When the Full option is selected, the program calculates the new window and then redraws all objects in the workfile. This option is very useful if part of a drawing has been "lost" as a result of repeatedly changing windows.

DELETE

The Delete option in the WINDOW Menu enables the designer to delete any previously saved window from the workfile. When this option is selected, the designer is required to enter the name of the window to be deleted from the workfile and hit RETURN. The program prints a message telling the designer that the window does not exist if it cannot find the window in the workfile. If it does find the window, it requires that the designer confirm the delete command before proceeding to eliminate the window.

GET

The Get option in the WINDOW Menu enables the designer to retrieve any window that previously has been saved in the workfile. When this option is selected, the designer enters the name of the window to be retrieved and hits RETURN. If the window does not exist in the workfile, the program prints a message telling the designer that it could not find the window. If the window is in the workfile, the program immediately redraws the current drawing using the selected window.

SAVE

The Save option in the WINDOW Menu enables the designer to save the window through which the designer is currently viewing a drawing. Windows are saved in the workfile. When this option is selected, the designer is required to provide the window with a name. The program does not permit saving two windows with the same name. If a window with the designated name already exists in the workfile, the designer must decide whether or not to replace it with the new window. When the program successfully saves a window, it prints a confirming message.

WHAT

The What option in the WINDOW Menu provides the designer with the absolute real-world limits of the current viewing window (Figure 8.9).

LIST

Several of the options in the WINDOW Menu require the designer to provide the name of a saved window. The List option in the WINDOW Menu allows the designer to list all of the windows by name that have been saved in the workfile. This option is quite useful when the name of a previously saved window cannot be remembered.

REDRAW

The Redraw option in the WINDOW Menu allows the designer to force the program to redraw all of the objects in the workfile at the current window.

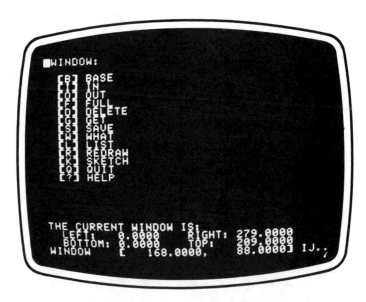

Figure 8.9 Using the What option—the coordinates of the current viewing window are displayed near the bottom of the screen.

ABSOLUTE

Some programs allow the designer to define a window by typing in the actual real-world coordinates of the desired window. When an option like this is selected, the program requires the designer to enter the real-world (i.e., "absolute") coordinates for three of the four edges of the desired window. The fourth edge is calculated automatically to preserve the correct window proportions. The program then uses this window to redraw the screen.

9

ADVANCED FUNCTIONS OF 2D SYSTEMS: SYMBOL LIBRARIES, AUTOMATIC CROSSHATCHING, AND INQUIRE

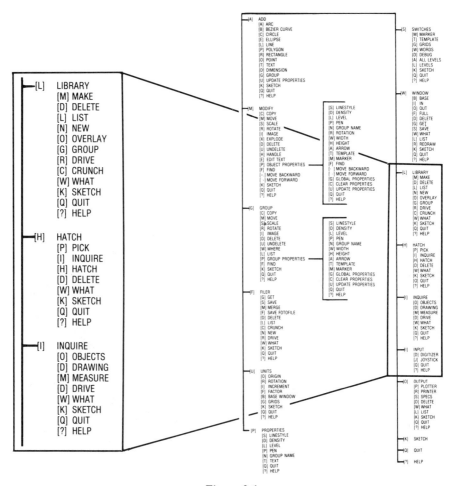

Figure 9.1

Many 2D CAD systems have a variety of very sophisticated options that enable the designer to perform functions with the computer that manually would be time consuming and/or tedious. As a rule, these functions require repetitive operations or considerable calculation. It is in the use of an automated drafting program for these functions and operations that CAD contributes directly to increased productivity. Three such operations that are commonly available in many microcomputer-based 2D CAD systems are described in detail in this chapter: symbol libraries, automatic crosshatching, and inquiry.

LIBRARY

A symbol library is a collection of symbols designed to meet the special needs of a unique set of drawings. Prepared libraries can be purchased from vendors in the same way that other microcomputer applications programs can be purchased, or they can be designed and created by the CAD user. Although symbol libraries may be created and accessed in different ways in different systems, the purpose of their use is always the same.

In the sample 2D program, each symbol library can have a maximum of 100 symbols in it; and the symbols are numbered from 1 to 100, starting from left to right and bottom to top in the overlay pattern used for their creation. The LIBRARY Menu in the sample 2D program allows the designer to make, delete, and otherwise manipulate symbol libraries.

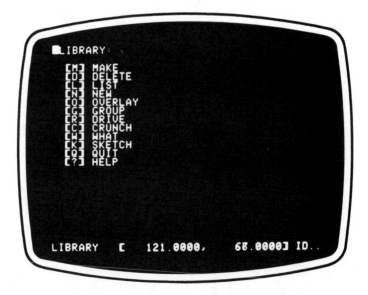

Figure 9.2 The (Symbol) LIBRARY Menu.

MAKE

The Make option in the LIBRARY Menu allows the designer to create a new, empty symbol library. When this option is selected, the program prompts the designer to enter the name of the new library. The program then creates the symbol library on the appropriate disk and prints a confirming message when finished.

DELETE

The Delete option in the LIBRARY Menu allows the designer to delete an entire symbol library. When this option is selected, the program requires the designer to enter the name of the library to be deleted. Once the name is entered, the program searches for that library, and when found requires that the designer confirm the Delete command. Once the command has been confirmed, the program deletes the library.

LIST

The List option in the LIBRARY Menu enables the designer to list the names, together with some other important information, about all of the symbol libraries that are stored on diskette, or on a hard disk.

The information displayed includes the disk name, the names of the symbol libraries stored on that disk, the last date that each symbol library was updated, the total number of symbols defined in the symbol library, the total number of primitive objects needed to define all of the symbols in the symbol library, and the percentage of the symbol library space that is currently used.

NEW

While creating a drawing, it is only possible to add a symbol from one library at a time. This library is known as the "active" library.

The New option in the LIBRARY Menu allows the designer to change the library from which symbols are currently being added to another symbol library.

When this option is selected, the program displays the name of the current symbol library and requires the designer to identify the name of the new library. The designer then attaches the appropriate menu matrix—a 10 by 10 plotted overlay of the symbols in the library—to the digitizer tablet and defines the lower left-hand corner and the upper right-hand corner of the menu for the program. The active library has now been changed, and the program displays a confirming message.

OVERLAY

The symbol library overlay is a 10 by 10 matrix of boxes, with each box containing one symbol from the symbol library. This overlay is designed to be placed on the surface of the digitizer pad and allows the designer to select a symbol while in the ADD Menu. The Overlay option is used by the designer to create a plotter ready overlay "menu" for the selected symbol library. When this option is selected, the program draws the 10 by 10 matrix of symbols in the "active" symbol library on the screen.

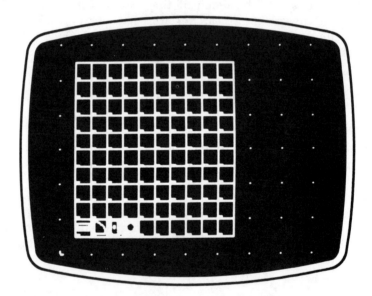

Figure 9.3 A Symbol Library 10 X 10 matrix displayed on the graphics screen. When the Library is completed, it is plotted and then taped onto the surface of the digitizer. Symbols can then be "picked" from the Library and placed in the drawing.

GROUP

A symbol is a user-defined grouping of objects. Each symbol (or group) is created by first making a drawing file that contains the collection of objects to be made into a symbol. This symbol drawing file is then added to the library. Symbols, themselves, are actually added to a drawing using the Group option in the ADD Menu. The Group option in the LIBRARY

Menu allows the designer to add, delete, or otherwise manipulate the individual symbols stored in a library.

The Add option in the Group submenu enables the designer to add individual symbols (or groups) to an already existing symbol library. Adding a symbol to a library basically amounts to copying the symbol from the 2D file into the library. Once added to the library, the original file containing the symbol can be deleted without affecting the copy in the symbol library.

When this option is selected, the program requires the designer to enter the name and number—remember 1 to 100—of the symbol to be added to the library. The number determines where on the symbol overlay the program draws the symbol.

The Delete option in the Group submenu allows the designer to delete an individual symbol from the current symbol library. When this option is selected, the program requires the designer to enter the number of the symbol to be deleted. When the number is entered, the program deletes the symbol and displays a confirming message.

The Handle option in the Group submenu allows the designer to change the handle point of a symbol in the current symbol library. When this option is selected, the program requires the designer to identify the symbol number and then to determine whether to use the "snap" feature when defining the symbol. The snap feature here is very similar to grid and increment snap. If enabled, it pulls the handle point to the nearest endpoint of the nearest line segment. This guarantees that the handle point will lie exactly on an endpoint.

The List option in the Group submenu lists the names and numbers of all of the symbols stored in the current symbol library.

DRIVE

Most of the options on the LIBRARY Menu make use of the symbol library diskette. For example, the Make option always makes a new library on that diskette. Similarly, the List and Delete options need to know about the symbol library diskette.

The Drive option allows the designer to change the symbol library disk drive where the 2D symbol libraries are stored. This option is useful if four or more disk drives are being used and the designer wants to list the symbol libraries located on any two of them.

CRUNCH

The process of adding and deleting symbols to a symbol library can leave empty (and unusable) "holes" in the library. The Crunch option allows the designer to "crunch" a symbol library and to remove those holes. This option should be used only when the program displays a message indicating that there is no more room in the library to add any symbols. Crunching the library may create the needed room.

WHAT

In systems that use four or more disk drives, libraries can be stored in a variety of places. The What option enables the designer to verify that the program is searching for a particular library on the correct drive. When the designer selects the What option, the program displays the current library name and the default library disk drive number.

HATCH

The HATCH Menu in the sample 2D program provides a quick and easy method for automatically creating equally spaced parallel lines on any angle within any area bordered by objects and text. The HATCH Menu also includes an area calculation feature. Applications for using this option include:

- *Mechanical drafting*: the production of section lines for an object's cut or broken surface
- *Architectural drafting*: the crosshatching of walls, patios, and other sectional views

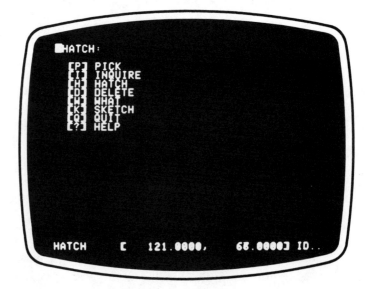

Figure 9.4 The HATCH Menu.

Other disciplines where there is a need to fill an area of a drawing with a solid color (or shading) can make use of this option by setting the spacing of hatch lines very closely.

The HATCH Menu is shown in Figure 9.4. Before hatching can be achieved, the designer must identify for the program the specific area of the drawing to use. The area is defined as a space on the drawing that is bordered by various objects (e.g., lines, rectangles, circles, and others).

In order to produce hatch lines the designer must:

1. Select the Pick option from the HATCH Menu. Pick an object that borders the area to be hatched. When all objects that border the area to be hatched have been "picked," the program calculates the exact borders of the hatch area.

2. Select the Inquire option from the HATCH Menu. The program then continually blinks all of the border line segments that surround the area to be hatched. This blinking provides a visual verification for the designer that the program is prepared to hatch the correct areas of the drawing.

3. Now select the Hatch option itself from the HATCH Menu. Enter the correct hatch parameters: spacing, angle, and hatch number. The program now inserts the correctly spaced parallel lines into the specified area of the drawing.

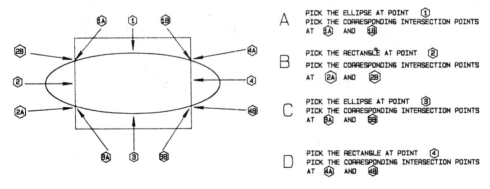

A   PICK THE ELLIPSE AT POINT ①
    PICK THE CORRESPONDING INTERSECTION POINTS
    AT ①A AND ①B

B   PICK THE RECTANGLE AT POINT ②
    PICK THE CORRESPONDING INTERSECTION POINTS
    AT ②A AND ②B

C   PICK THE ELLIPSE AT POINT ③
    PICK THE CORRESPONDING INTERSECTION POINTS
    AT ③A AND ③B

D   PICK THE RECTANGLE AT POINT ④
    PICK THE CORRESPONDING INTERSECTION POINTS
    AT ④A AND ④B

Figure 9.5 Using HATCH—picking the area to be hatched.

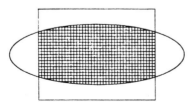

Figure 9.6 Using HATCH—the completed crosshatch.

Regardless of the 2D program, the Hatch operation is performed in a similar fashion.

PICK

The Pick option in the HATCH Menu allows the designer to define the borders that surround the area of the drawing to be hatched. When this option is selected, the program requires the designer to use the cursor to identify the object to hatch. Then it is necessary to identify all objects—or parts of objects—that border the object to be hatched.

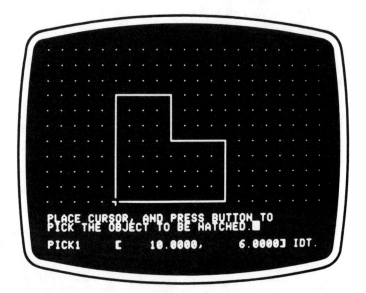

Figure 9.7 Using the Pick option in the HATCH Menu.

If an object only partially borders on the object to be hatched, it is necessary for the designer to identify the intersection points for the program. When all the objects that border the hatch area have been picked, and the Pick option is terminated by the designer, the program then calculates the exact border of the hatch area.

INQUIRE

Hatch is one of the more difficult operations performed by the 2D CAD program. The designer uses the Pick option to define the borders that surround the area to be hatched. The Inquire option enables the designer to verify that the boundaries the program created for the hatch area are indeed correct.

When the designer selects the Inquire option, the program will continually blink each border segment of the hatch area. To stop the blinking of the boundary line segments and to return to the HATCH Menu, the designer need only hit the RETURN key. If the hatch area was not properly defined using the Pick option, the program displays an appropriate error message.

HATCH

The Hatch option in the HATCH Menu actually enables the designer to insert equally spaced parallel lines on any angle into a predefined area of the drawing. When this option is selected, the program displays the input screen as shown in Figure 9.8. The program positions the designer to enter each of the hatch parameters in the order shown.

Figure 9.8 Using the Hatch option in the HATCH Menu.

The hatch spacing parameter refers to the perpendicular distance between any two parallel hatch lines. Initially, it is set to be the value of the smallest increment at that time. The designer, however, can enter any real-world coordinate for the spacing.

The hatch shift parameter allows the designer to create double-line hatching on the drawing. To create double-line hatching, the designer enters any real-world coordinate for the shift. The hatch shift is initially set to 0.0000.

The hatch angle parameter allows the designer to specify the precise angle (in degrees) at which the hatch lines are to be created. The angle is measured from the horizontal with a counterclockwise direction. The hatch

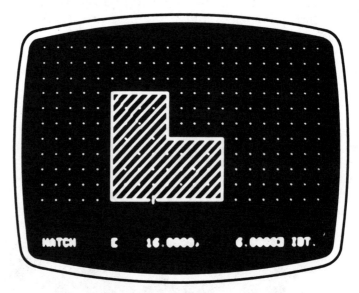

Figure 9.9 A single hatch at 45°.

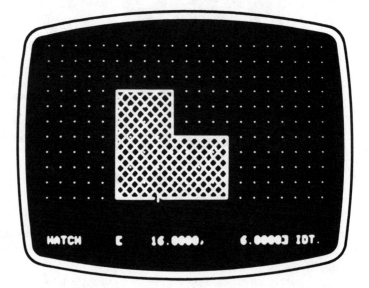

Figure 9.10 A double hatch, at 45° and 135°.

angle parameter is initially set to 45 degrees, but the designer can enter any angle within the range -360.0000 to 360.0000.

The hatch number parameter allows the designer to identify the hatch lines to be created with a specific number. This is useful when it is necessary to use the Delete or What option from the HATCH Menu. The hatch number is initially set to 1, but the designer can enter any number up to 32,767.

Once all required parameters have been entered, the program begins inserting lines into the hatch area.

DELETE

The Delete option in the HATCH Menu allows the designer to delete a set of hatch lines by their hatch number. When this option is selected, the program prompts the designer to enter the hatch number. When the number is entered, the program searches for all hatch lines with that number and deletes them.

WHAT

The What option in the HATCH Menu allows the designer to inquire where a particular set of hatch lines are in the drawing. When this option is selected, the program prompts the designer to enter the hatch number. When the number is entered, the program searches for all hatch lines with that number and then blinks them once.

AREA

Some other 2D programs are able to perform additional functions in the creation of hatch lines beyond that of the sample program. An Area or similar option calculates the approximate area and perimeter of any bordered area. To use this type of program option, the designer:

1. Defines the objects that border the desired area
2. Verifies that the correct border lines have been selected
3. Selects the area option

The program then requires the designer to determine the number of divisions to be used. Often a program might "suggest" a particular number of divisions, but the designer can require a larger number for more accuracy. The program then calculates the area and perimeter enclosed by the borders and displays the results.

INQUIRE

Most 2D CAD systems provide one or more options that enable the designer to obtain valuable specific information about the nature of a particular drawing. The location of the beginning and ending points of a line, the radius of a circle, the distance between two points, the area and perimeter of a rectangle, which levels are being used in a particular drawing, and much more might be specific information a designer would find helpful in continuing with a drawing or in using that drawing as part of another more complex one.

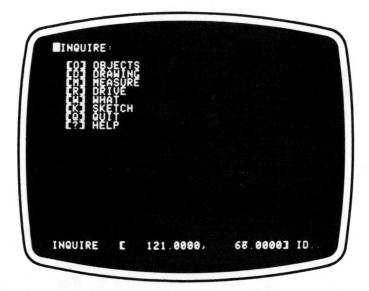

Figure 9.11 The INQUIRE Menu.

The options in the INQUIRE Menu of the sample program provide the designer with access to this type of information about a drawing. Figure 9.11 shows the INQUIRE Menu. All of the options off the INQUIRE Menu display information in a specific format and each option has a specific purpose:

1. If the designer requires specific information about an object on a drawing (such as initial position, rotation, length, or area), the designer needs to select the Objects option.
2. If the designer requires information concerning how many of the available levels are currently being used or how many plotter pens are necessary to plot, the designer needs to select the Drawing option.

3. If the designer needs specific measurements on a drawing (such as area, length, or rotation), the designer needs to select the Measure option.

Different systems may provide this type of information in a somewhat different format, but the purpose of accessing is always the same—to get additional needed information about an aspect of a drawing.

OBJECTS

The Objects option in the INQUIRE Menu of the sample program displays the various geometric characteristics of a selected object. When this option is selected by the designer, the program searches for the last object on the drawing. The program then displays a screen as shown in Figure 9.12. This display reflects the specific geometric characteristics of the object, a circle: center point, beginning point, radius and diameter, and circumference and area. Such information can be used by the designer at a later time. For example, the designer might want to place a second object, perhaps an arc's endpoint, *precisely* on the endpoint of the displayed object. The information provided enables the designer to use the Absolute input mode to directly enter the arc's coordinates.

The geometric characteristics of other objects (e.g., polygons, text, rectangles, etc.) are all different. When a particular object is selected, the

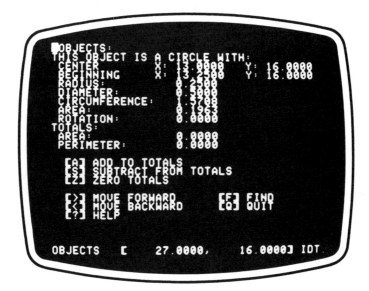

Figure 9.12 Using the Objects option to display the physical and geometric characteristics of a Circle.

program automatically updates the display to reflect the proper characteristics. Other characteristics might include beginning and ending points of lines and other primitive objects.

In Objects, all coordinate points displayed, and measurements made, are expressed in "real-world coordinates." Rotation is always expressed in degrees and measured counterclockwise from the horizontal axis.

The Objects option in the INQUIRE Menu provides a number of functions that assist the designer in obtaining information about a drawing:

Add: causes the program to add the current object's area and perimeter to the totals field.

Subtract: causes the program to subtract the current object's area and perimeter from the totals field.

Zero: causes the program to clear ("zero") the totals field.

Find: permits the designer to select any object on a drawing by picking it with the cursor. This option is identical to the Find option in the MODIFY Menu. Once the object is found, the program updates the geometric characteristics.

Move Backward: allows the designer to sequentially move backward through the list of objects in a drawing. This option is identical to the Move Backward option in the MODIFY Menu. Once the object is found, the program updates the geometric characteristics.

Move Forward: allows the designer to move forward sequentially through the list of objects in a drawing. This option is identical to the Move Forward option in the MODIFY Menu. Once the object is found, the program updates the geometric characteristics.

DRAWING

The Drawing option in the INQUIRE Menu displays a summary of the levels and plotter pens used by a drawing. When this option is selected, the designer must provide the name of a previously saved drawing file that the designer wants to view. The program then searches through this drawing and displays a matrix indicating which levels of all of the possible levels are being used in the drawing. It also displays the plotter pens that have been designated for use (Figure 9.13).

MEASURE

The Measure option in the INQUIRE Menu allows the designer to measure the distance, angle, and area between any two arbitrary points on a drawing. When this option is selected by the designer, the program prompts the

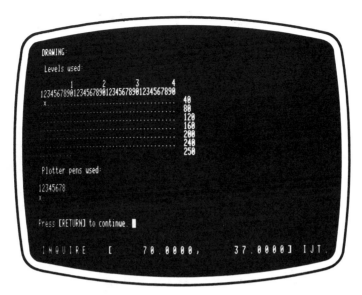

Figure 9.13 Using the Drawing option to display levels and pens used in a drawing.

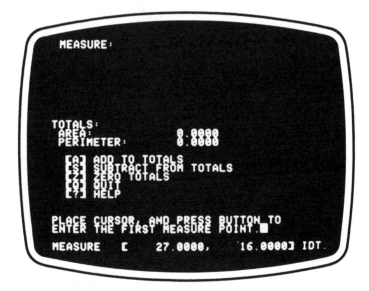

Figure 9.14 Using the Measure option in the INQUIRE Menu.

designer to identify the first point on the drawing to be measured. The designer moves the cursor to the desired point and then presses the cursor button. Next the program prompts the designer to define the second point on the drawing to be measured. Again, the designer moves the cursor to the

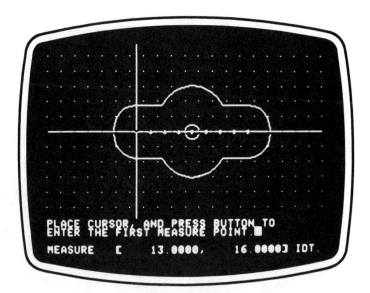

Figure 9.15 Using the Measure option—entering the first measure point.

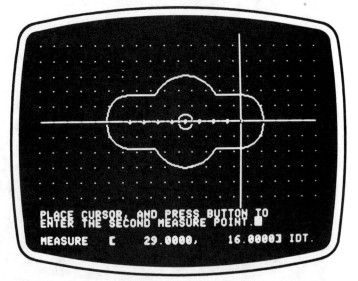

Figure 9.16 Using the Measure option—entering the second measure point.

desired point and then presses the cursor button. After both points have been defined the program displays the information shown in Figure 9.17.

The sample program provides several options to aid the designer—several are similar to those available in the Object option:

Add: causes the program to add the current area and length to the totals field.

Subtract: causes the program to subtract the current area and length from the totals field.

Zero: causes the program to clear ("zero") the totals field.

Measure: permits the designer to enter two arbitrary points on a drawing and causes the program to display the distance, angle, and area measurements.

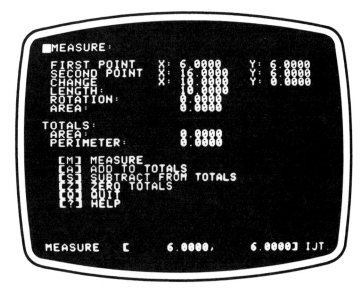

Figure 9.17 Using the Measure option—displaying the information obtained.

DRIVE

The Drive option in the INQUIRE Menu permits the designer to override the default data drive number for systems employing either a hard disk or three or more disk drives.

WHAT

The What option in the INQUIRE Menu displays the current default data drive number. This option is used by the designer to verify that he or she is, indeed, searching for a drawing file on the correct drive number.

10

2D EXAMPLES: MECHANICAL, ELECTRICAL, AND ARCHITECTURAL

A 2D general purpose-drafting system can be productively employed in any number of disciplines, including mechanical, electrical (and electronic), and architectural drafting and design.

The productivity gain from the use of 2D general-purpose drafting derives from using past work to shorten the time required to do future work. In each discipline, that means creating standard symbol libraries, or parts of drawings or completed drawings, that can be modified and assembled for future work.

This chapter presents some examples of products from various disciplines with a brief description of how each drawing was accomplished. The utilization of the functions, features, and operations of a typical low-cost 2D CAD system, as discussed in the previous chapters, is illustrated in each drawing below.

MECHANICAL DRAFTING

Guidepost

Figure 10.1 is a detailed drawing of a saw fixture used by the University of California's Prosthetic Devices Research Center. The drawing points out computer drafting capabilities of linestyles, pen weights, lettering, and computer-generated title block and border. The system can also be used with preprinted forms.

148

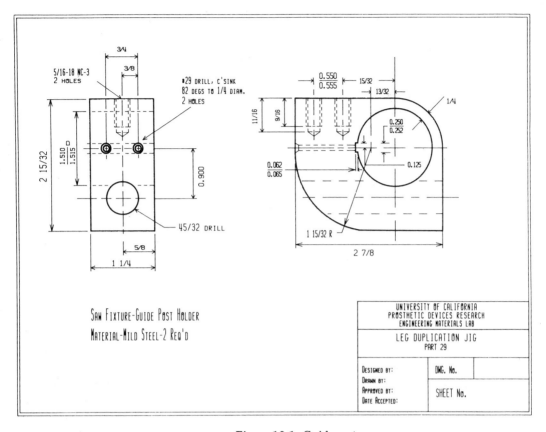

Figure 10.1 Guidepost.

The fixture drawing was prepared using a 2D general-purpose drafting package. The drawing was then plotted on an 11 inch × 17 inch plotter. Here is how the 2D general-purpose drafting system was used to prepare the drawing:

First, the designer prepared the title block and border and saved the title block on the diskette. (The same title block format can then be used over and over again for subsequent drawings.)

Second, the designer used the zoom or window capability of the system to focus on one portion of the drawing, expanding it to fill the entire screen. Then the designer proceeded to draw the one view of the fixture by alternately selecting the lines, arches, and other figures as appropriate.

Using an automatic dimensioning feature, the designer touched two points on the screen, and the system automatically prepared the witness lines and automatically calculated the distance. Then, text was added by placing the location of the text with the cursor and then typing in the text using the keyboard. In most 2D systems the designer would be working interactively with the system, responding to its prompts or cues.

When the right-hand side of the drawing was completed, the designer returned to the original screen or window and viewed the entire drawing in proportion. At this point, any errors that were detected or modifications that were desired could be made before proceeding.

The designer then proceeded to window in on the left-hand side of the screen, creating that view of the fixture in a manner similar to that described for the first view.

When both views were completed to the satisfaction of the designer, the entire drawing was merged with the title block previously saved to diskette. The completed design was then drawn on a high-resolution plotter and sent to the engineer for checking.

The benefits of using CAD to prepare mechanical drawings come from saving the part drawn and using it again as a component of a larger drawing or as the starting point for future designs. Many people claim productivity advantages of 2:1 or better when compared to manual drafting as well as increased consistency of quality.

Manifold

A second mechanical drafting example, a manifold, features crosshatching and an exploded detail. See Figure 10.2. In the drawing of the manifold, the designer started by setting the screen size to proportions representative of the actual size of the manifold. To facilitate placement and precision, a 10×10 grid was then established. (Grids are used to aid in the precise placement of objects on the graphics screen and are not drawn when the drawing is plotted.)

A variety of options in what is usually known as the "Add" function or menu were used to draw the basic outline of the manifold and the details were then filled in with centerlines. A crosshatching option was used to automatically calculate the location of and then to draw all of the cross-hatch lines.

Next the designer retrieved a standard border previously saved on diskette and the manifold was scaled to fit within the border (not shown). This scaling action is usually accomplished through a function known as "Group," where all of the lines were added under the same "Group" name so that this scaling could be accomplished.

Then the manifold was dimensioned using an autodimensioning feature. In some systems the dimension text can be set to feet and inches; in others it must be manually edited using the keyboard.

The designer next zoomed or windowed in to create the exploded view. First, the area to be exploded was identified and then included in a window to fill the entire screen. Next, the group name of all the objects was modified

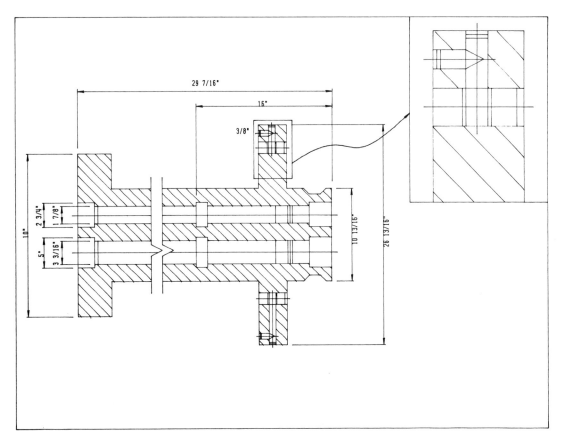

Figure 10.2 Manifold.

and a common group name established. Then, after returning to the base or original window, a copy was made of the group representing the area to be exploded and the copy was scaled "up" and moved to the location of the exploded area.

A Bezier curve with an arrow on the end was used to point from the main body of the manifold to the exploded view.

After adding the break lines and other details, the manifold drawing was made automatically on the plotter.

Details

Sheets of detail (Figure 10.3) can be built up from separate drawing files either by merging several files in the workfile or by separately calling up and plotting the details on the same sheet.

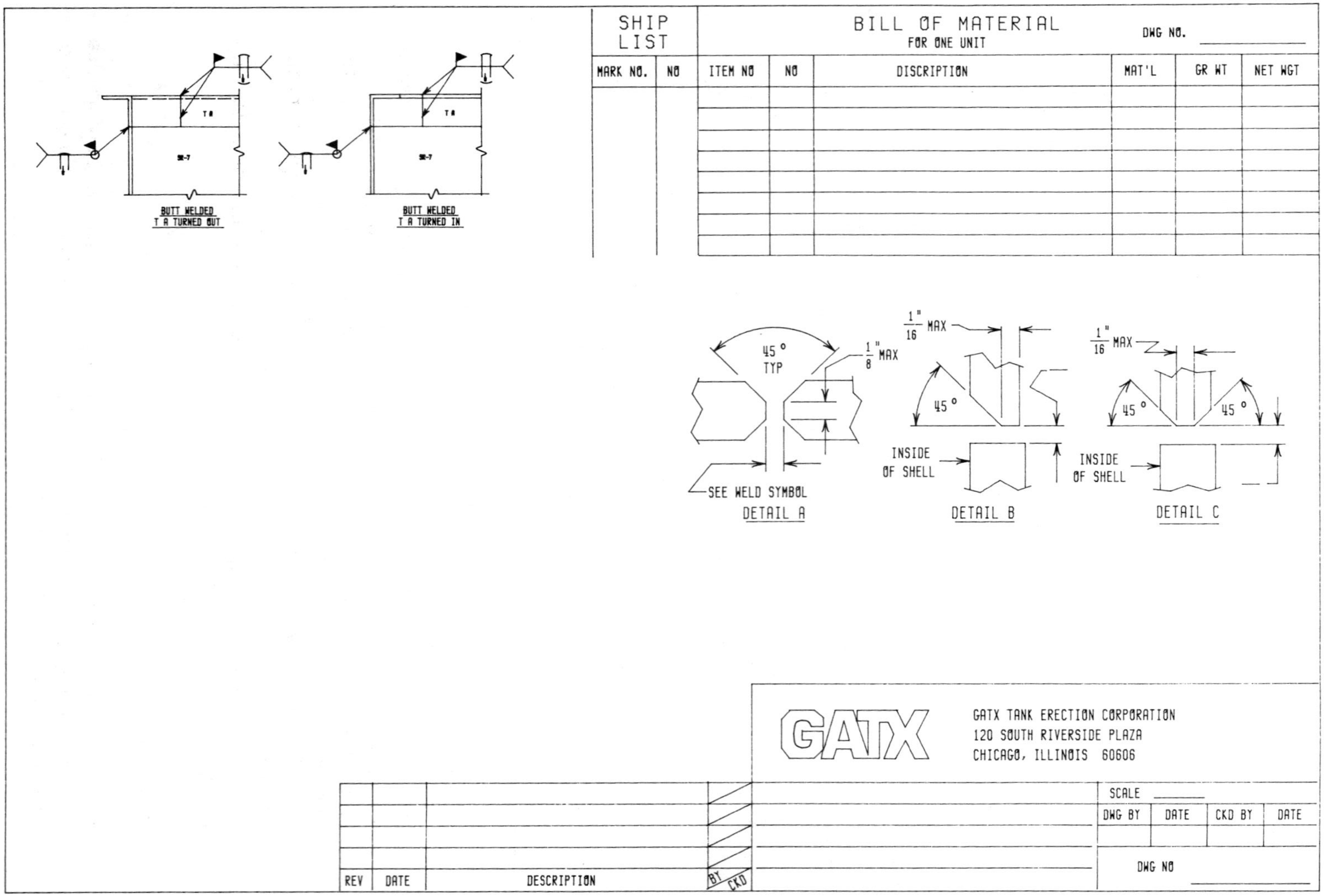

Figure 10.3 Details.

To use the workfile approach, the designer first "Gets" or retrieves the border and title block previously saved on diskette. Then the designer retrieves the actual drawing, in this instance "BUTT WELD T A TURNED OUT," and as a single group locates it in the upper left-hand corner of the drawing. (A "Move" or similar option in the "Group" menu is used to accomplish this.) Next, the designer retrieves "DETAIL A" and moves it to its proper location. The procedure is the same with each of the other details. Then the designer sends the merged workfile to the plotter.

The other possible approach is for the designer first to call up the border, send it to the plotter, then to clear the screen (and workfile) and to retrieve "DETAIL A." DETAIL A is then moved to its proper location and plotted. Again, the workfile and screen are cleared, DETAIL B is retrieved, located, and plotted. This procedure is continued in this fashion until the drawing is completed.

The advantage of merging the workfile as described in the first option is that the entire drawing can be viewed on the screen before sending the drawing to the plotter. The advantage of the second option, merging on the plotter, is that the designer is working with a smaller workfile and the system responds more quickly.

ELECTRICAL DRAFTING

Ladder Diagram

Control systems or instrumentation schematics have long been a productive application of CAD technology. For the ladder diagram in Figure 10.4, the designer created one circuit, scaled that down, and then used a group copy option to very quickly replicate that circuit five times. Next, a modify option was used to edit the minor differences between each group. Finally, the drawing was sent to the plotter for rapid preparation of the final product.

Logic Diagram

Using computers to prepare digital logic diagrams (Figure 10.5) was one of the first applications of CAD.

Standard symbols and title block are created once and then used over and over. Use of X and Y axis lock (i.e., where lines and other objects are "locked" in parallel with the X or Y axis) facilitates the drawing of the lines. Text entry, editing, and adjusting provides standard high-quality text.

Printed Circuit Board Outline

The outline of the printed circuit board (Figure 10.6) provides an example of where mechanical drafting and electrical drafting merge.

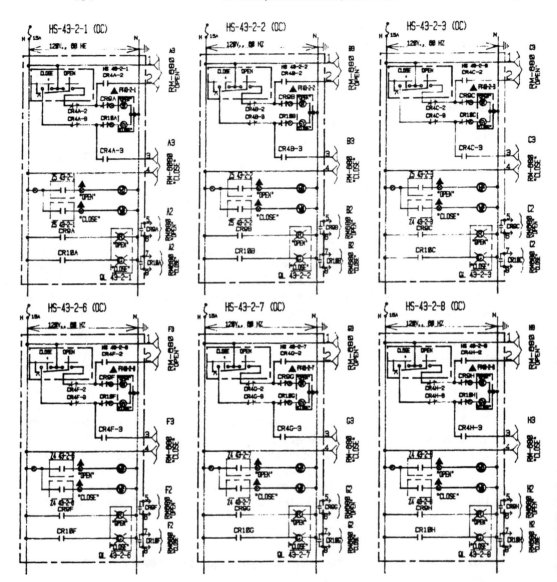

Figure 10.4   Ladder diagram.

ARCHITECTURAL DRAFTING*

In utilizing the system for an architectural project there is one element that becomes common to each step along the way. That element is the need to develop a standardized approach to architectural drafting. A standardized approach greatly enhances the success of using the system. The following

*This section was developed by Carl Walls.

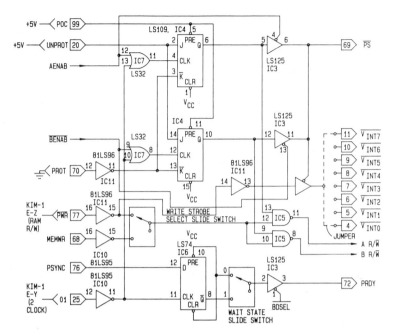

Figure 10.5 Logic diagram.

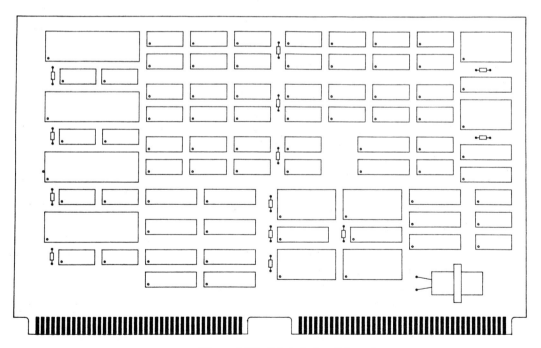

Figure 10.6 Printed circuit board.

narrative is intended to help explain some of the steps taken in developing a sample architectural construction drawing project. The steps and approach are in no way intended to represent the only way, but to help the design student develop the best way.

The sample project illustrated for architectural drafting is a motel project which utilized multiple generation of basic floor plan types and repetitive symbols. The sample is a partial project showing examples of unit plans, building plans, elevations, and schedules. The individual drawings that make up each sheet were saved on diskettes as separate drawings that were recalled in any order to make up a sheet. This helped in editing and arranging the drawings on the final sheet.

The architect started the project by first "translating" the actual sheet size to the computer coordinates. This can be accomplished by thinking of each sheet size in relation to the scale of the drawing to be done on it. If the plan is to draw 1/8-inch plans, the size of a 24 inch × 36 inch sheet becomes 192 × 288—multiply the sheet size by the scale per inch. This then means that each 1.000 computer coordinate becomes 1′0″ at the scale chosen. These coordinates are then used to set the "base window," or the original full screen of the drawing.

Once the computer coordinates of the base window were established, the architect proceeded to develop the sheet border and title block. Once the border layouts were constructed, they were saved onto diskette for use in future projects.

Next, the architect proceeded to create the architectural, electrical, and plumbing symbols for this project. (Many of these symbols are common to other projects and can be saved as symbol libraries to be retrieved for use in other drawings.)

For ease of identification and use, the architect placed a copy of each symbol on a portion of the sheet layout where they could be copied, moved, or otherwise manipulated for use in the drawing. Each symbol should have its own group name to aid in its use.

Many systems have a feature that enables the designer to create a mirror image of a complicated or detailed object. Such a mirror image function was used by the architect to create these symbols. Half of the symbol was created by the architect and then the computer was used to generate a mirror image of the other half of the symbol.

Several important decisions related to the characteristics of a drawing, including the level of different features (most 2D systems allow the designer to work on as many as 36, 64, or even 250 or more levels) have to be made for each drawing. This architect kept a log for each new drawing in which the level and group name of individual elements were identified as well as a brief description of the group.

Levels are the layers that designers put together to form a drawing.

They enable the designer to work on a portion of a drawing without incurring the extended redraw time of a complete drawing when the screen is refreshed in windowing in or out. As an example, the dimensions on the sample project are on different levels from the symbols and all of these are on different levels from the notes. Using levels is a great aid on larger and more complex detailed drawings. Another example of using layers is in the room finish schedule. Here the basic schedule layout can be on one level, while the items that would change from project to project would be on a different level, allowing changes to be made easily and quickly.

Most 2D systems give the designer a great deal of flexibility in the use of scaling of drawings. When using a drawing of one scale to create a new scaled drawing, the need for reduction is eliminated by transferring the original scaled drawing to a sheet established with the new scale. This process was used by the architect to create the 1/8-inch scale building plans in the sample. The basic floor plans were done at 1/4-inch scale and transferred to the 1/8-inch scaled sheet layout. Care must be taken in the transfer of text which is usually set in relationship to the base window established at the outset by the designer. Individual sets of text can be modified in turn or all text can be transferred as a single group and then modified as one. This can be achieved by using the "Group Properties" or similar function of any 2D system where the individual properties of an object (including text) such as height or width, linestyle, level, or pen can be modified by the designer to fit the particular needs of the drawing.

The "Group" menu is handy for making multiple images, reversing images and rotating objects in drawings, as well as moving and deleting groups of objects in any drawing. The architect used a group function to make the constructing of the building plans and elevations as well as the reverse basic plans easy and accurate. When copying and moving groups, it is necessary to take care that all levels are checked so that the transfer of only the desired groups is accomplished.

A "Template" function is useful for laying out drawings and guidelines. Templates disappear when changing windows—or in zooming in or out—or when a drawing is plotted. Templates should be deleted or removed from a drawing when they are no longer useful in preserving workspace in the workfile.

All drawings used in these architectural samples were located on the sheet in relation to where they would actually appear on the final layout. This was accomplished by determining the coordinates of each drawing based on the scale of the sheet and then using those coordinates in the base window function. This was not required but did enable the architect to speed up the final sheet assembly by eliminating the need to move complex drawings across the sheet. Because each drawing was created independently of each other, the combining of drawings was much easier.

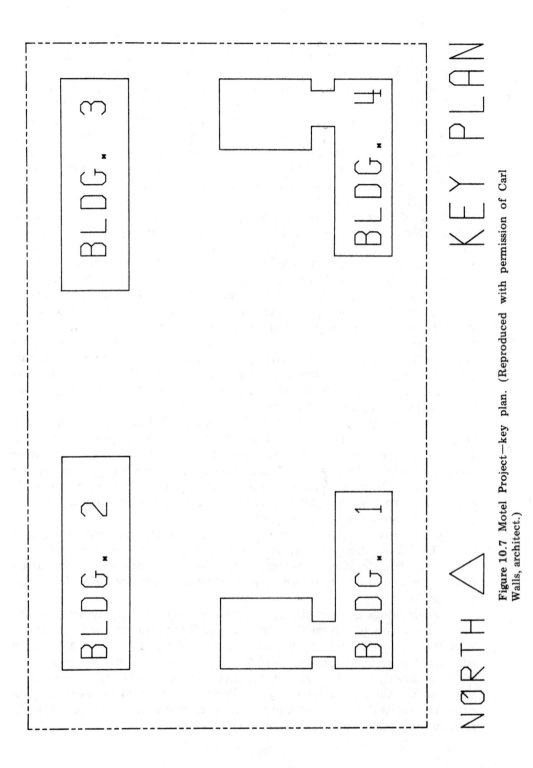

Figure 10.7 Motel Project—key plan. (Reproduced with permission of Carl Walls, architect.)

158

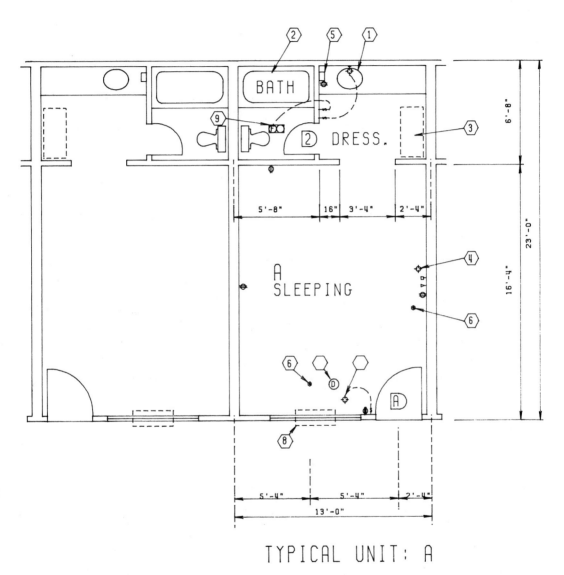

Figure 10.8 Motel Project—typical unit: A. (Reproduced with permission of Carl Walls, architect.)

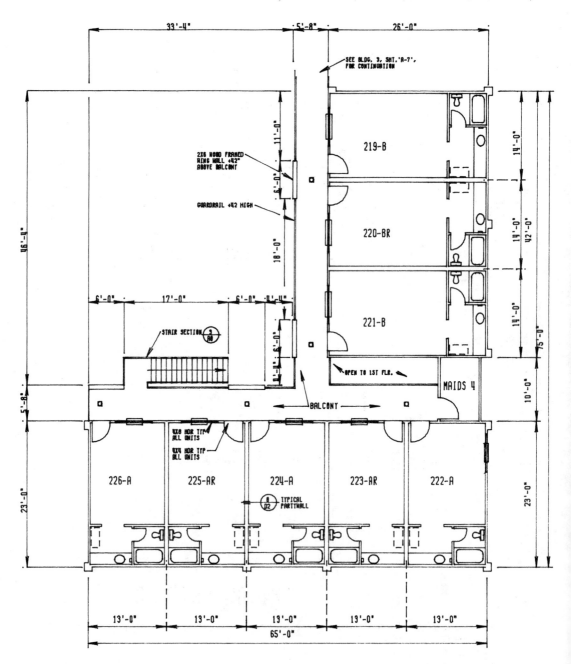

BLDG. 4: SECOND FLOOR PLAN

Figure 10.9 Motel Project—building 4: second floor plan. (Reproduced with permission of Carl Walls, architect.)

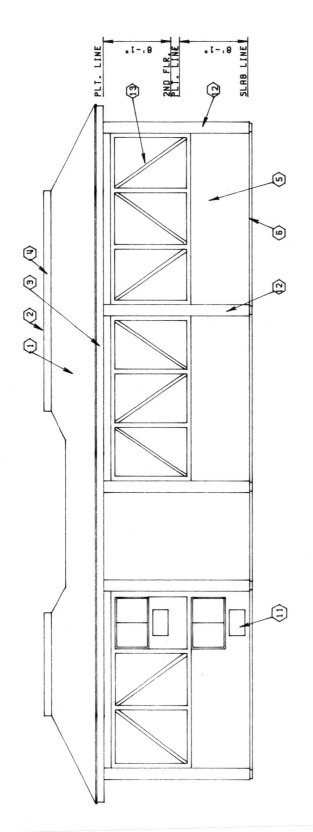

Figure 10.10 Motel Project—building 4: east elevation. (Reproduced with permission of Carl Walls, architect.)

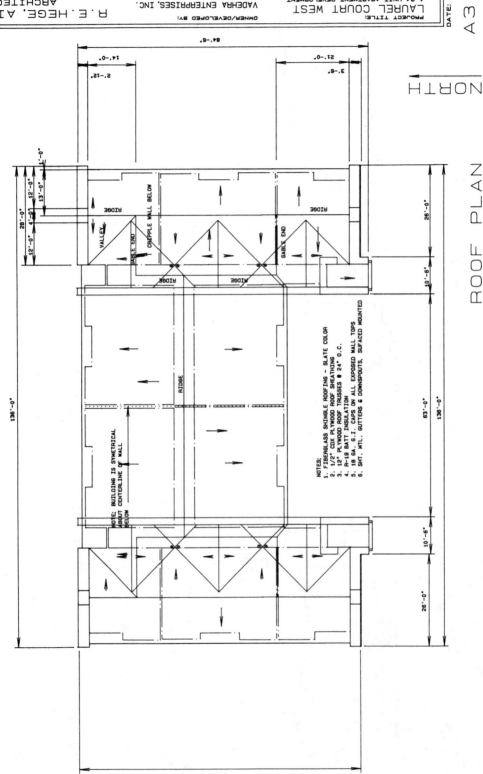

ROOF PLAN

Figure 10.11 Motel Project—roof plan. (Reproduced with permission of Carl Walls, architect.)

NOTES:
1. FIBERGLASS SHINGLE ROOFING - SLATE COLOR
2. 1/2" CDX PLYWOOD ROOF SHEATHING
3. 12" PLYWOOD ROOF TRUSSES @ 24" O.C.
4. R-19 BATT INSULATION
5. 18 GA. G.I. CAPS ON ALL EXPOSED WALL TOPS
6. SHT. MTL. GUTTERS & DOWNSPOUTS, SUFACED MOUNTED

NOTE: BUILDING IS SYMETRICAL ABOUT CENTERLINE OF WALL BELOW

NORTH

RIDGE

VALLEY

GABLE END

CRIPPLE WALL BELOW

PROJECT TITLE:
LAUREL COURT WEST
A 24 UNIT APARTMENT DEVELOPMENT
7704 LAUREL CANYON BLVD.
LOS ANGELES, CALIFORNIA

OWNER/DEVELOPED BY:
VADEHRA ENTERPRISES, INC.
8464 SUNSET BLVD, SUITE 736
LOS ANGELES, CA 90028
(213) 466-7666

R.E.HEGE, AIA
ARCHITECT
6356 GREEN VALLEY CIR. #311
CULVER CITY, CA 90230
(213) 410-0866

DATE:
SHEET NO:
A3.3

ROOM FINISH SCHEDULE

ROOM NAME	FLOOR	BASE			WALLS		CEILING			NOTES	MATERIALS & FINISHES
	MAT	MAT	FIN	HGT	MAT	FIN	MAT	FIN	HEIGHT		
TYPICAL UNITS :											MATERIALS----------
SLEEPING	A	E	-	-	F	1	F	3	8'-0"	R1	A...CARPET
DRESSING ROOM	A	E	-	-	F	1	F	3	8'-0"	R1 & R2	B...RESILIENT TILE
BATHROOM	B	E	-	-	G	2	F	2	8'-0"	R1 & R2	C...
											D...
											E...TOP-SET VINYL
											F...1/2" STD. GYPBD.
MAIDS / LAUNDRY:											G...1/2" W.R. GYPBD.
MAIDS	B	E	-	-	F	1	F	3	8'-0"	R1	H...
LAUNDRY	B	E	-	-	F	2	F	2	8'-0"	R1 & R2	J...EXPOSED CONCRETE
ELECTRICAL	J	E	-	-	F	4	F	4	8'-0"	R1	
											FINISHES-----------
											1...FLAT WALL PAINT
											2...SEMI-GLOSS ENAMEL
											3...SIM. ACOUSTIC TEXT.
											4...TAPE & FILL ONLY

R1...REFER TO 1/8" BLDG. PLANS FOR PARTWALL LOCATIONS.

R2...PROVIDE RESILIENT CLIPS AT CEILINGS BELOW HARD FLRS. ABOVE.

Figure 10.12 Motel Project—room finish schedule. (Reproduced with permission of Carl Walls, architect.)

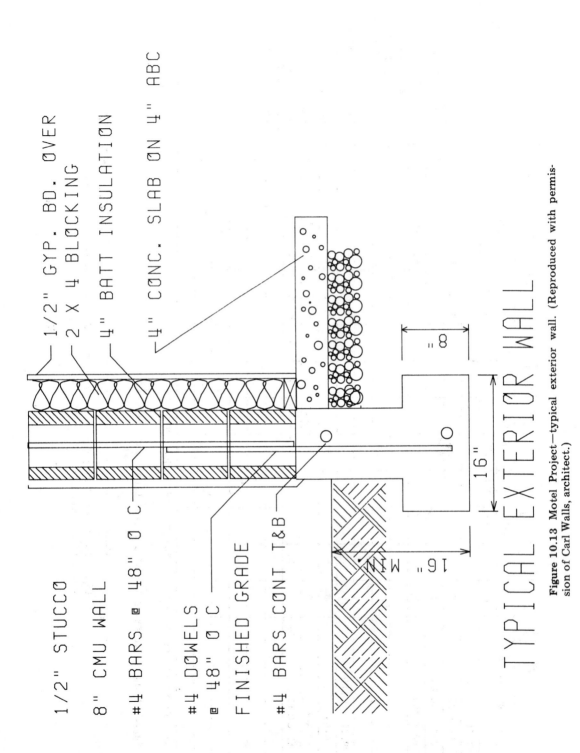

1/2" GYP. BD. OVER
2 X 4 BLOCKING

4" BATT INSULATION

4" CONC. SLAB ON 4" ABC

1/2" STUCCO

8" CMU WALL

#4 BARS @ 48" 0 C

#4 DOWELS
@ 48" 0 C

FINISHED GRADE

#4 BARS CONT T&B

8"

16"

16" MIN.

TYPICAL EXTERIOR WALL

Figure 10.13 Motel Project—typical exterior wall. (Reproduced with permission of Carl Walls, architect.)

SYMBOL LIBRARIES

Libraries of symbols can be used over and over again to increase productivity in certain types of drafting. Figures 10.14 and 10.15 show the libraries for architectural drafting and HVAC drafting.

ARCH-L1

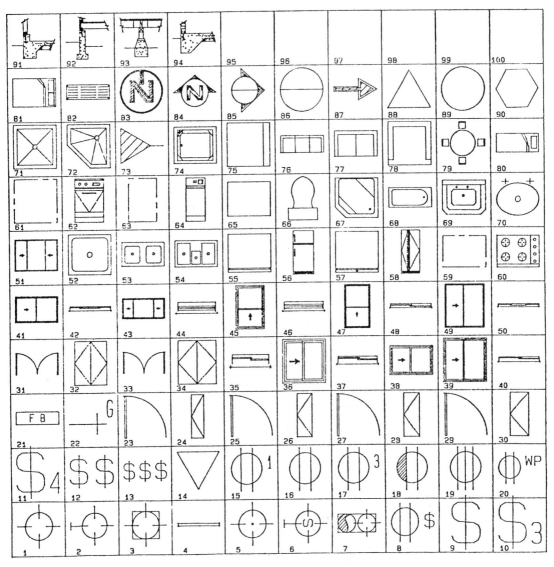

Figure 10.14 Symbol library for architectural drafting.

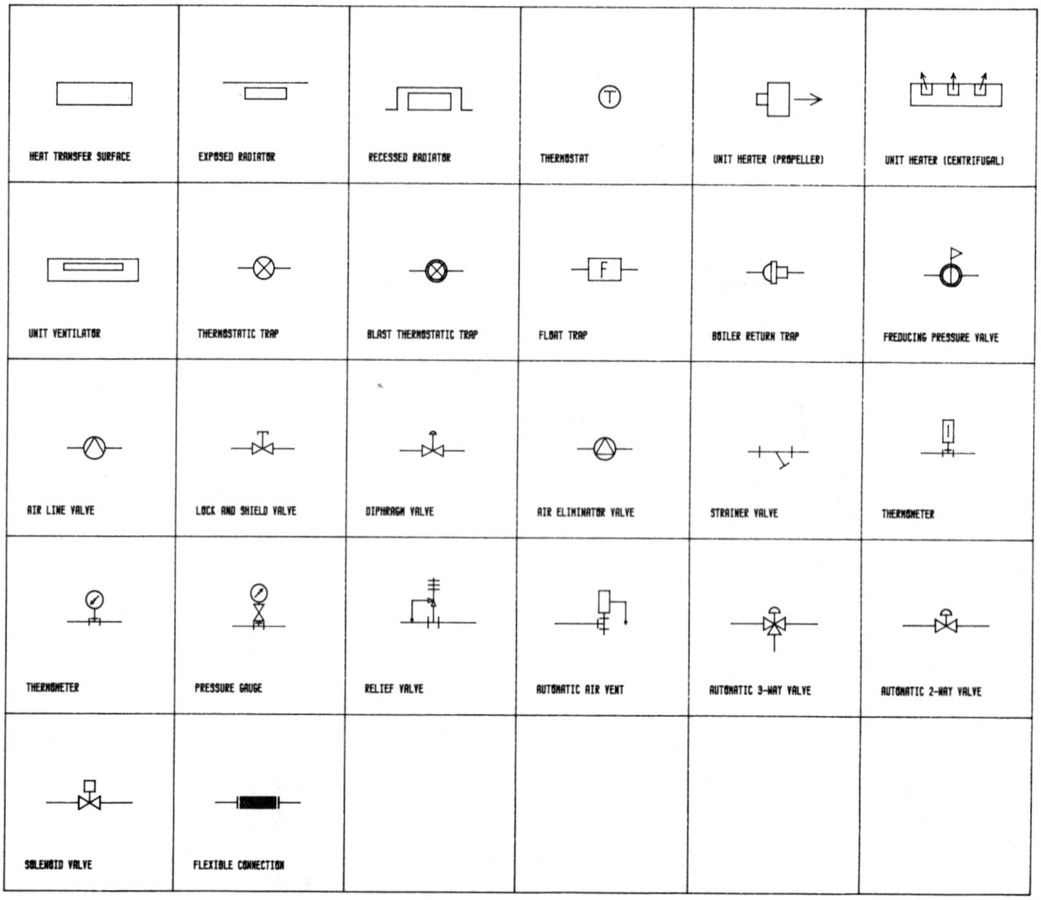

Figure 10.15 Symbol Library for HVAC.

A beginning library is supplied with most computer programs and the designer is able to modify or extend the supplied libraries or to create entirely new ones. Libraries are created or modified using the general-purpose drafting system just like creating or modifying any other drawing. Libraries are collected together using a built-in command of the system and each individual figure in the library is then assigned to a location in a table much like the one that follows.

To use the library, the designer first plots the symbol library to the scale of the digitizer pad, then tapes the plotted table over the digitizer pad. It is then possible simply to pick the desired symbol off the pad and place it at the desired location on the graphics screen.

11

3D DESIGN SYSTEMS

A 3D design system is used to create a computer-based mathematical model of a physical object using the keyboard to input the coordinate points and to manipulate the orientation of the object. A general-purpose 3D system can be from one of three categories: solid modeling, surface modeling, or wire frame. In solid modeling, the objects are described as physical solids. In surface modeling, objects are described as empty containers, with a surface but no internals. In wire-frame modeling, 3D objects are described as edges only, like a birdcage. This chapter will focus on one general-purpose 3D system that uses the surface modeling technique. As a further restriction, the sample program used here is limited to 3D objects with surfaces made of plane polygons.

A typical 3D system contains the following basic hardware components:

1. The display monitor (the CRT or screen)
2. The keyboard
3. The boxlike device that houses the floppy disk
4. Another boxlike device that houses the central processor

The display monitor in the system shown is used to display both text and graphics. Other systems use one monitor for graphics—sometimes color, usually high resolution—and a second monitor for the text. The keyboard is used for entering information into the computer. After input, the information is stored on a removable recording medium called a floppy disk or diskette. Essentially, the components operate in the same fashion for 3D as they do for the 2D system.

The sample 3D program featured in this chapter accepts input to define an object made of polygons, provides for viewing that object or a collection of objects from a scene, and then provides for transferring the resultant view to the 2D component. The 2D program is then used to annotate, dimension, plot, or otherwise manipulate the view in 2D style. The sample 3D program has the following features:

1. The program is menu driven.
2. The program displays the options that can be selected with a single keystroke.
3. The program is interactive.
4. The program has "default" actions for easy startup, but the designer can change the defaults as desired.
5. The program provides a true perspective display of a scene composed of 3D objects.
6. The 3D display can be a static view or it can be a dynamic view whereby the viewer's eye moves around the screen in a prescribed manner.
7. Lines that are hidden in a given view are removed and are not displayed.

INPUT TERMINOLOGY

As noted above, the 3D program receives input from the keyboard. In the 3D program, the word "cursor" refers to the marker that moves on the screen guiding the input of data. The input from the keyboard defines points

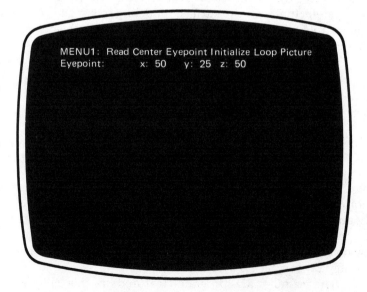

Figure 11.1 Displaying the distance from the Eyepoint.

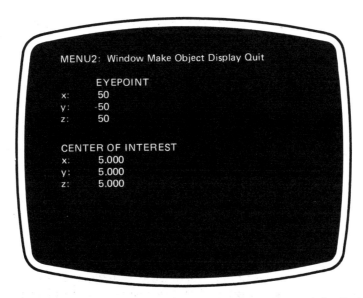

Figure 11.2 Displaying the distance from the Eyepoint and the Center of Interest.

in X, Y, Z coordinates that are the vertices of polygons that define the surfaces of the objects being created. As an object is placed in a scene it is located at a given X, Y, Z distance from the origin of the scene. This location is referred to as DELTA X, DELTA Y, DELTA Z.

Two points are necessary to be able to define a true perspective view, the "eyepoint" and the "center of interest." The "eyepoint" is the X, Y, Z location of the point in the scene that is being viewed. The "center of interest" is the focal point of the drawing—and the focal point from the eyepoint.

SYSTEM THEORY

The sample 3D program uses polygons to build more complex objects in 3D. To define objects made of polygons, the designer must list the polygons individually. This is done by listing the coordinates of each vertex (point) of the polygon in clockwise order (as seen from outside of the object) around the periphery of the polygon. It is important that all polygons be described consistently since the clockwise order is useful for determining which side of a polygon is facing the viewer.

A solid object is defined as a group of adjoining polygons. Since neighboring polygons share vertices along common borders, objects are defined by first listing all the vertices belonging to the object and then listing polygons by the number of vertices they use.

The designer wants the choice of viewing an object from any viewpoint. The input to determine a viewpoint requires two points in space: the position *from* which the designer is looking (the eyepoint) and the position *at* which the designer is looking (the center of interest).

The clipping window defines the field of vision in much the same way that the film gate and lens in a camera limit the field captured by the film. The window can be defined as a polygon corresponding to the boundaries of the display as the designer expects to view it. For example, if a 12-inch video display is viewed from a distance of about 16 inches, the clipping window should be a rectangle about 6 inches high and 8 inches across, located 16 inches from the eyepoint.

Any polygon found within the field of vision is displayed. It is necessary to achieve the appearance of perspective (i.e., objects in the distance should be smaller).

Hidden parts of polygons are removed from the display. Polygons are assumed to be convex. The vertices of all polygons are assumed to be defined in a clockwise fashion as seen from outside the object. Many objects are closed surfaces, meaning that the inside of the object can be seen only by passing through the surface. In fact, if we choose to do so, we can construct all objects as closed surfaces for display purposes.

In any event, if a polygon appears on the screen with its vertices in a counterclockwise order, it is being seen from the inside. If the designer is looking at a closed surface from the inside, some other part of the surface must lie between the designer and the polygons in question. Therefore, when making pictures of solid objects made of closed surfaces, the designer must immediately reject any polygon appearing in a counterclockwise order.

When the eyepoint lies on the positive side of the plane of a polygon, the vertices of that polygon appear in clockwise order. When on the negative side, they appear in counterclockwise order.

SYSTEM ORGANIZATION

The sample 3D program is organized around a collection of several "menus." A menu is a list of options or commands available to the designer. Each menu in the program serves a specific purpose. For example, PICTURE displays the current scene on the graphics screen, while READ brings another object into the scene.

All available options are displayed in the *two* MAIN Menus. The MAIN Menus are displayed automatically when the 3D program is first executed by the designer. The MAIN Menus allow the designer to reach the specific 3D functions that are needed to create drawings. Each option in the MAIN Menus has a key letter that is pressed to cause that option to be selected.

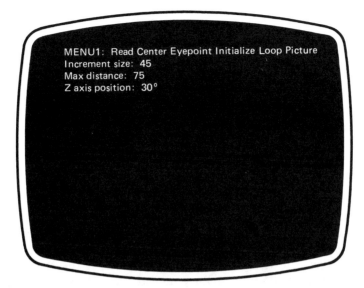

Figure 11.3 The options on Menu1—Read, Center, Eyepoint, Initialize, Loop, Picture.

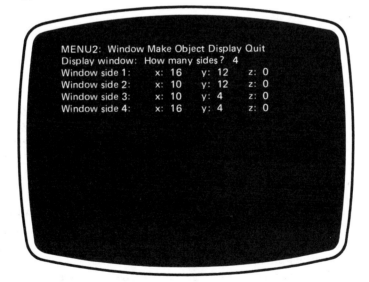

Figure 11.4 The options on Menu2—Window, Make Object, Display, Quit.

Read

The READ function allows the designer to read previously created information (or data) related to an object. When this function is selected, the program prompts the designer to enter the name of the previously saved file.

When the system finds that file, it asks the designer to enter the delta x, delta y, and delta z coordinates that define the placement of the object with respect to origin of the coordinate system of the scene being created.

Objects are defined in their own space or coordinate system. The delta x, delta y, and delta z are distance of the object with respect to the origin of the coordinate system being created. When the input values of the delta x, delta y, and delta z are entered, the program automatically returns to the MAIN Menu. A scene can be created by "reading" several objects or by "reading" the same object several times.

Center

The CENTER function allows the designer to input the coordinates of the center of interest—the point in the scene at which the designer is looking. The default values of the center of interest are the coordinates of the origin of the screen, usually: X=0, Y=0, Z=0.

When the designer selects this function, the system requires that the designer input delta x, delta y, and delta z. After these values are provided by the designer, the program automatically returns to the MAIN Menu.

Initialize

The INITIALIZE function initializes and sets default values of the key variables and also clears the screen.

The eyepoint is set at X=50.0, Y=−50.0, and Z=50.0. The center of interest is set at X=0, Y=0, Z=0. The four points defining the window are:

Window (1)	X=−4.0	Y=−3.0	Z=16
Window (2)	X=−4.0	Y= 3.0	Z=16
Window (3)	X= 4.0	Y= 3.0	Z=16
Window (4)	X= 4.0	Y=−3.0	Z=16

Loop

The LOOP function allows the designer to view an object by rotating it 360 degrees around the Z axis in specified increments. When this function is selected, the system prompts the designer to enter the increment size, the maximum distance, and the Z-axis position.

The increment size is the angle that the eyepoint should be stepped through. For example, setting the increment size to 30 degrees means that the program steps through the angle between the eye and the Z axis from 0 degrees to 30 degrees, 30 degrees to 60 degrees, and so on until the 360-degree rotation is completed.

Maximum distance is the radius from the Z axis to the eyepoint as the eyepoint revolves around the Z axis. The Z-axis position is the "elevation" above the scene.

Once the parameters above have been entered and accepted, the program displays the views of the object until the 360-degree rotation is completed. When the 360-degree loop is completed, the program automatically returns to the MAIN Menu.

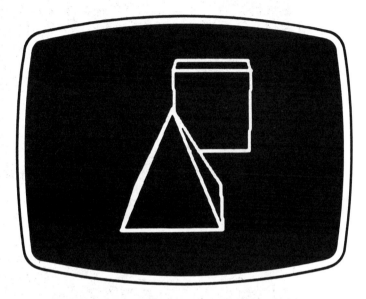

Figure 11.5 One moment in the continuous Loop.

Figure 11.6 A second moment in the continuous Loop.

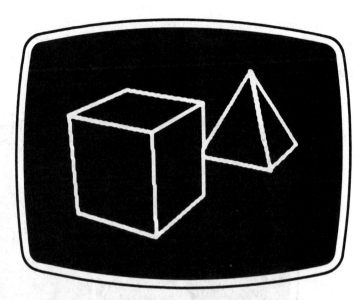

Figure 11.7 A third moment in the continuous Loop.

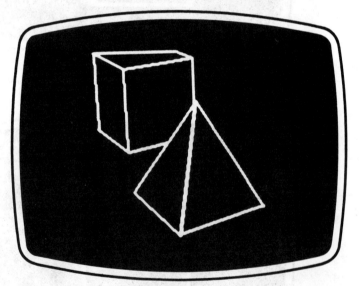

Figure 11.8 A fourth moment in the continuous Loop.

Picture

The PICTURE function draws the scene on the screen. If the picture does not appear on the screen, the designer must check that the coordinates of the eyepoint and the center of interest are correct. The scene contains the objects that have been read.

Window

The window is a surface through which the eye looks to the point of interest and on which the image is displayed. The default window is a 6 by 8 unit horizontal rectangle located 16 units above the origin of the scene.

Makeobject

The MAKEOBJECT function is used to create new objects, edit existing files or previously defined objects, and display data of any existing files containing object definitions. When the designer selects this function, the program displays two options: Edit and Display.

When the Edit option in the MAKEOBJECT Menu is selected, the program prompts the designer to enter the file name. The designer enters the file name and when the program finds it, it displays the additional set of options on the screen. Hitting RETURN signals the program that the designer wants to create a new file.

To insert the points of a polyhedron (box), the designer hits the "I" (Insert) key and then enters the points in any order. Points are entered in the left-handed coordinate system, with the positive directions of Z axis up, X axis right, and Y axis into the screen. Pressing "Q" signals the program that the last point has been entered.

If a mistake has been made, the designer may change the point numbers by moving the cursor to the number to be changed, pressing "C" (Change), and then typing in the correct number.

If it is necessary to remove a point completely from the list of points, the designer moves the cursor to any of the numbers of that point and then presses "R" (Remove).

When all of the numbers of all of the points have been correctly entered, the designer uses Quit to leave Point Edit and to enter Polygon Edit.

In Polygon Edit, the Insert command is used to add polygons. When polygons are added, they must be entered in a clockwise rotation when facing the outside of the polygon being entered.

The commands Change and Remove serve the same functions here as they did in the Point Edit option. When the designer is finished adding polygons, terminating the option causes the program to exit the Polygon Edit option and the program then prompts the designer to save the newly created file or the changes made to an existing file.

The Display option in the MAKEOBJECT Menu allows the designer first to examine the "List of Points" data. Hitting RETURN once after this option is selected causes the program to display the "List of Polygons" data. Hitting RETURN a second time returns the program to the MAKE-OBJECT Menu.

Output

The OUTPUT function allows the designer to save the information necessary to plot the screen image of the objects on a plotter. In the sample program, the saved file can be retrieved with the 2D component and can then be plotted, updated, dimensioned or otherwise enhanced.

Display

The DISPLAY option in the MAIN Menu allows the designer to review information about the x, y, z coordinates of the current Center of Interest. When the review is completed, hitting RETURN returns the program to the MAIN Menu.

12

3D SPECIAL PURPOSE: PIPING AS AN EXAMPLE

Systems that model three-dimensional objects can be either general purpose or special purpose. General-purpose 3D systems, such as the one discussed in Chapter 11, are useful for modeling 3D objects but are limited in terms of utility for a particular application.

Special-purpose 3D systems retain much of the 3D modeling characteristics but provide functions that specifically suit a particular application. This chapter focuses on a sample special-purpose 3D system, one called T-SQUARE, which was developed to facilitate piping design, detailing, drafting, and material takeoff. Much of this chapter is drawn, by permission, from an article that appeared in *Plant Engineering*.

Designers who manually develop pipeline sketches from layout drawings or a physical model are all too familiar with the tedious and time-consuming cleanup steps they must perform after the creative process requiring skilled judgment is completed. These cleanup steps are required to make design drawings acceptable for permanent entry into the company's drawing system and to produce the documentation required to order, cut, and install the parts of the pipeline. Computer-based piping layout systems aid the designer by eliminating the drudgery of these cleanup steps, automatically calculating dimensions, producing the final high-quality isometric drawings, and generating bills of materials. Now designers can perform these cleanup steps at the touch of a button, instead of spending hours going through the pipeline sketches, manually accumulating like items, and looking up catalog and part numbers—as well as having to remember all relevant company standards to produce the written bill of materials. Examining a new computer-based piping drafting system, T-Square, from T&W Systems, Inc., illustrates how such systems enhance the design process.

The heart of the T-Square system is the Terak 8510, a graphic computer system consisting of a 16-bit microcomputer, two floppy disks, a keyboard, and a CRT video monitor. The basic system is complemented by a digitizer tablet, a pen-type plotter, and a complete turnkey software package for creating piping isometrics.

The designer uses the digitizer tablet to create a pipeline sketch that is displayed on the CRT video monitor. The digitizer is a flat tablet, approximately 11 inches $\times$ 11 inches, with a special pen attached. Wherever the user touches the pen on the surface of the tablet, the pen position is electronically converted to digital coordinates and transmitted to the computer. In the T-Square system, the digitizer is used to select piping items such as ELLs and TEEs from a "menu."

To use the menu to select an ELL, the next element in the pipeline, for example, the user touches the pen over the symbol of an ELL. Once the sketch has been entered into the computer with the digitizer, the designer can edit it until he or she is satisfied that it is correct. Then, with a few simple commands, the computer draws the final pipeline diagram on the attached electronic pen plotter in just a few minutes.

The pen plotter produces a fine draftsman-quality drawing on command from the computer. The plotter draws lines, symbols, numbers, and characters. The plotter uses replaceable pens that accommodate various media, including bond, vellum, and Mylar. It produces drawings up to a maximum size of 11 inches $\times$ 17 inches.

The designer uses the keyboard to enter textual input such as pipeline descriptions and lists of standard item descriptions that are used on the bill of materials. On command, the Terak computer can count up all parts and print a complete bill of materials.

USING THE SYSTEM

The process of using the system begins with the preparation of the Piping Specifications and Piping Description files and storing them on the floppy disks. The Piping Specifications are the company's standards for piping items for a particular project. They consist of a list of the part numbers ("codes") to be used for each type of item for each type of service. For example, part number 376J41 might be the correct selection for all 300-psi water service ELLs between 1.5-inch and 4.0-inch diameters. The part numbers are defined by the user and may represent catalog numbers, internal part of purchasing spec numbers, or any other numbering system. Then if the user later selects an ELL in a run of 3-inch pipe while building a pipeline, the computer will automatically select the correct part number, 376J41, for that ELL when constructing the bill of materials for that pipeline. The Piping Description file consists of the company's standard written description for each part defined in the specification file. For example, part number 54543 may be given the standard description "PIPE SCH40SMLSA53-B." The standard descriptions are also used by the computer when it constructs the bill of materials. The Specification and Description files save the designer all of the "looking-up" work and time. They also assure consistency and adherence to company standards by all members of the design team.

The Specification and Description files are entered into the T-Square system via the keyboard and are stored on the floppy disks. The piping supervisor or project manager may also choose to distribute only a subset of the standard Specifications to each member of the team, thus assuring that only certain classes of parts are available for use in certain sections of the project.

In most organizations, after completion of the initial effort required to enter the company's specifications and standard item descriptions into the T-Square system, the only ongoing effort required is periodic updating of these files to keep them current.

PREPARING PIPING ISOMETRICS

The individual designer can begin the actual design process after he or she has received a set of floppy disks with the authorized Specifications and Descriptions. The first step is to construct the assigned pipelines. With the T-Square system, there are two parts to this process. First the designer enters the "topology"—the way the parts are put together. Then he or she inputs the "coordinates" or geometry of the pipeline. From these two sets of data the computer constructs a scaled isometric drawing of the pipeline. The designer uses the digitizer menu to enter both types of data. The user simply starts at one end of the pipeline, touching the symbol on the digitizer menu for each item in the sequence in which they are to be connected. The user also inputs pipe size, ASA pressure class, and other definitions for each item in the pipeline. Next, the designer uses the menu to input geometric information ("coordinates") required to define the pipeline's location in space (e.g., the ELL is 5'0" EAST of the GASKET and at a centerline elevation of 110 feet 0 inches). Coordinates may be expressed in absolute (plant coordinates), relative, or polar coordinates. "True Angles" (fitting angles) may also be specified for custom bends in either two or three dimensions.

Once the topology and all coordinates have been entered, the T-Square system can be instructed to construct a picture of the defined pipeline and display it on the CRT. The system performs all the calculations required to convert relative, polar, and "true angle" coordinates to absolute coordinate values, scales the picture to fit on the display screen (and on the plotter), and displays it on the CRT. It also calculates and displays all dimensions. No calculations are required by the designer in order to produce the display. The system assumes certain orientations for the symbols of items in the pipeline, and automatically places the dimensions on the drawing. At this time, the designer may modify the orientations and positions of these elements to achieve a satisfactory presentation. The designer may also scale segments of the pipeline up or down, or zoom in on a segment for detailed inspection. The T-Square system automatically performs all calculations for these operations. When the designer is satisfied with the isometric drawing, he or she can instruct the T-Square system to plot it on the plotter.

BILL OF MATERIALS

The T-Square system produces a bill of materials for each pipeline. On command, it selects the correct part number for each part used from the Specifications file, adds up the quantities of each part, and retrieves the part description from the Description file. If it is provided with the dimensions of the various standard parts (flange to flange dimension of a valve, for example), it can also compute the correct cut lengths of all the pipe pieces and include these in the bill of materials.

The bill of materials can be plotted or printed separately or included on the same drawing. The bill of materials can also be transmitted to another material control system computer if desired, using T-Square's telecommunication capability.

COST SAVINGS

Beyond relieving the drudgery of some aspect of piping layout design, computer-based systems afford higher productivity and significant cost savings. T&W Systems reports that detail designers using their system show a 2:1 improvement in productivity. Hourly overhead for a T-Square system is about $3.00, based on the assumption that the $20,000 system will be used 40 hours per week for five years. Viewed this way, the T-Square system provides the same productivity as a second detail designer, at only an additional $3.00 per hour. T&W estimates that the system pays for itself after about 200 isometrics.

CONCLUSION

Special purpose 3D systems can offer considerable productivity improvement in the specific discipline in which they are employed. The advantage of the 3D special-purpose system is that the terminology and functions are all specific to the user and consequently maximize the user's productivity. The disadvantage is that the cost of development can be returned by use only in the discipline for which it is intended. Consequently, the application will not pay out unless there is a large volume of work of that specific discipline.

13

SELECTING A CAD SYSTEM

A CAD system represents a tremendous potential for increasing productivity for any enterprise involved in design or manufacturing. However, at the same time, if left to chance, the CAD system may represent an expensive mistake that adds little, if anything, to productivity and competitiveness. The best way to ensure that a CAD system is a wise investment is to employ a well-proven methodology for determining the specific requirements of the industrial, manufacturing, or design enterprise; for identifying potential solutions, and then selecting the alternative that best meets the enterprise's unique set of requirements.

Although there are a number of methodologies presently in use, the SPECTRUM methodology is used in this chapter to provide one possible guide for the evaluation and selection of a CAD system.

WHY A METHODOLOGY

A good methodology such as SPECTRUM can make a significant contribution in solving some all-too-common problems associated with computer system development or selection. These common problems include schedule slippages, cost overruns, user dissatisfaction, poor documentation, and low management confidence. A methodology can help in a variety of ways.

Problems of schedule slippage and cost overruns are caused by poor estimating techniques, "ballooning" scope, poor visibility during a project, and overwork of team members. A methodology contributes to a solution by providing estimating guidelines, emphasis on capturing accurate user requirements, design freezes, and simple status reporting.

Users of a CAD system are sometimes dissatisfied with the system chosen. When dissatisfaction sets in, among the primary reasons is that no effective method of capturing requirements was employed, or no quality control process was used, or the systems group forced a solution, or there was a lack of strong user commitment. A good methodology provides a procedure for capturing user requirements and an organized way for defining and reviewing the final decisions.

A consistent methodology can produce other positive results: documentation standards, quality review points—in short, a standard and logical approach to a project. Use of a methodology provides some assurance that if the steps are followed, a successful project outcome will result.

OVERVIEW OF THE METHODOLOGY IN SELECTING CAD

The role of a methodology is to assure that the analyst selecting a CAD system knows how to do the work, that is, to collect user requirements, design a system, perform a feasibility study, make a selection, and then plan for and install the system.

The backbone of any good methodology is the project life cycle, that is, the tasks that must be executed to complete the project of selection. SPECTRUM provides several hundred tasks for a development life cycle and about 100 for a project designed to select a packaged CAD system. Each task in a project life cycle is composed of three to five pages of description of how to do the task, a definition of the tangible results, forms to use to elicit responses and evaluations, and estimating guidelines. The life cycle controls the sequence of tasks; and results from one task are required before starting the next. Tasks are grouped into project phases, with reviews specified at key points.

The sections below provide an example of a likely CAD system user and then follows the SPECTRUM methodology for selecting a hypothetical CAD system.

ABC ARCHITECTS AND ENGINEERS

ABC, our example firm, employs eight professionals—only three are full time—who spend at least 50% of their time on a drawing board either designing or drafting. A rough time analysis indicates that about 10% of that time is used drawing 3D sketches such as perspectives and isometric views during the design phase. The major portion of their time, however, is involved in layout, detailing, drafting, and making schedules and bills of materials.

ABC is prospering and the company is facing an increasing work load. The management has decided that they would like to install a CAD system, starting with one station and expanding to four or five as they become more accustomed to the CAD system and as their work load continues to grow. They would like to have the potential to expand to 20 stations over the next two to five years.

These are the steps that ABC Architects and Engineers might follow for selecting a CAD system using the SPECTRUM methodology.

Tasks in the Selection Process

The first step in this, as in any project, is to organize it. In this instance, initial organization requires assigning a person or team of practitioners to make the analysis and establishing a management group to review the analysis, weigh the recommendations, and make decisions. Although each company might be different, each management team, nonetheless, should be well represented by prospective users of the system.

The next step in the project is for the analyst or analysts to write a proposal for the project. The proposal should clearly identify the purpose of the project and should provide an estimate of time and cost for carrying out the project, as well as the estimated total cost of purchasing and installing the selected CAD system. A sample proposal form is included at the end of this chapter.

It is important for both the analyst and for management to realize that the estimate of total costs provided in the proposal is only a rough approximation, since the actual scope of the project has not yet been developed. The estimate of total project cost, however, is a necessary part of the proposal because it helps management establish an appropriate priority for the project in relationship to other capital-intensive projects. To the best of his or her ability, the analyst should provide accurate estimates of the time required and the anticipated cost for the next phase of the project, but the analyst should be held accountable for only that estimate.

This is sometimes called the "moving window" of estimating. An estimate of total costs to the end of the project is required for planning, but the analyst—or analyst team—is held accountable only for that part of the project that is well defined.

The next step in the systems project to select a CAD system is to determine user requirements. First, the analysts should collect example drawings, design procedures, and other documents that help to define how the company does its work. These documents form the basis for the development of a questionnaire that the analysts can use to interview a sampling of management as well as the designers and drafters who will actually use the system once it is purchased and installed. The interviews should address improvements required in the firms operations and are documented on the questionnaires. These needs are then summarized by the analysts and then compiled into a requirements narrative which also lists those that should be reviewed with the interviewees and management.

The review process that follows is one of the important aspects of an orderly methodology for selecting a CAD system. The purpose is to involve all the parties in a realistic incremental way so that the selected system will be well accepted by all.

After the requirements of a CAD system are determined and reviewed, an initial list of desired functions and features is developed by the analysts. This list of desirable functions and features is then used as the primary basis for selecting among the various alternative CAD systems that will be evaluated. Extended tables indicative of the types of functions and features that might be required of a CAD system are included in the Appendix.

Looking back at the needs of ABC Architects and Engineers, several needs emerge clearly. The CAD system selected must be modular so that ABC can start with one station and add additional stations as their growth and work load requires. ABC will need a system that does some 3D design work and is especially good at 2D drafting, since the bulk of their work is 2D. Also, the selected system must be capable of being used to prepare bills of materials and schedules. And it would be desirable if the system had the ability to automatically generate these bills of material.

User friendliness and the relative ease of learning the CAD system would always be important considerations. Time spent learning the system is time not spent being productive on the job. At ABC user friendliness would be an important consideration because many of the potential users are part-time employee/professionals.

With only these criteria established, an initial list can be developed to screen candidate systems. The list in Table 13.1 might well be an initial list developed by the analysts for ABC.

Screening values can be applied to each of the criteria identified in the extended CAD Hardware and Software Evaluations in the Appendix. The screening value is used to eliminate candidate systems from further consideration and thereby to narrow down the selection options. For example, in the chart just above, the modularity criteria, with a weight of 100, would eliminate any CAD system that cannot be started with just one station. The networking criteria, with a weight of 90, would dictate that it must be possible to network up to 20 stations if the growth forecasted by ABC is to be provided for.

All of these criteria are objectively set by the analyst team before even previewing a demonstration of a CAD system; and all of them are based on the requirements that were developed earlier and which were reviewed and approved by both management and potential system users.

TABLE 13.1 SCREENING CRITERIA

Criteria	Weight (1-100)	Screening value	Option A	B	C	D
Modularity	100	Able to start with 1				
Networking	90	Up to 20 stations				
User friendly	90	Learn in 2 days				
2D capability	90	90% of desired features				
Multidiscipline	90	Architectural, mechanical, electrical				
3D capability	80	Perspective and hidden line removal				
Bill of materials	60	Some desirable				
Cost	60	Less than $30,000 per station				
Screen resolution	60	400 × 300, minimum				
Screen size	60	At least 12 inches diagonal				
Response time	60	1/2 second between commands; repaint 500 objects in 15 seconds				
Capacity	60	At least 2000 objects per drawing				
Plotter	60	At least 20 ips, 0.001 resolution, 6 pens				

The step of identifying candidate systems can be undertaken concurrently with the development of the screening criteria. The best method here is to research the available systems through one of the reference services, such as DataPro, Data Sources, or DataTech. DataTech might be an especially good resource since they are devoted exclusively to research in CAD systems.

ABC would find perhaps 50 CAD systems listed in the DataTech guide. The analyst would start two lists for the rough screening, one headed "Systems Screened Out" and the other "Systems Under Consideration." Then the analyst might start from the first system listed in the DataTech guide, and would determine which of the two lists that system—and each system in turn—would be placed on. If a system were screened, a reason would also be noted. For example, a system might be screened out with the notation "over $30,000 per station" or "won't network over eight stations."

Probably, a relatively short list of "Systems Under Consideration" would then emerge—probably no more than four or six names. At this point a new screening criteria list would be created, one that draws on information provided in the guide for detailing the values of each system under consideration on the short list.

Before proceeding, another review meeting should be held with management. All the options must be considered and all the systems should be reviewed and fully discussed, including those already screened out. This review meeting allows every advocate of a particular system to be heard before the project progresses to a point where it is difficult to turn back. It also provides an opportunity for any hard feelings to be set aside and for a consensus to emerge—at least on the validity of the process, if not on the narrowing of the choices. Remember, seeking input and evolving consensus are strategies that will ensure project success. The ultimate result of this review meeting should be approval of a short list of systems to be viewed and evaluated in much greater detail.

The next step in the process is to verify the values assigned to each criterion for each vendor. The best approach is to see a demonstration and then to contact companies that already have that particular system in operation. Questions should be asked about the performance and reliability of the hardware and the software, other specific technical features of the system, the service that can be expected from the vendor, and any other questions that might help in the evaluation of the candidate systems. At this point the field usually narrows down to two or three systems that appear best able to meet the needs of the company. Here the weightings that were originally established for the criteria can be halpful. Also, it may be helpful for the two or three finalists to come in and to make a full presentation to the analyst team, and even to the full management review team.

If there are still several vendors on the short list, a formal proposal procedure can be employed. The user requirements developed earlier plus the evaluation criteria can be included in a formal "request for proposal" (RFP). The RFP can then be sent to each of the vendors on the short list, requesting each to detail how their system can meet the requirements and criteria of the RFP. The proposals should be reviewed by the analyst who should be responsible for developing a recommendation for awarding the winner in the proposal process. The recommendation and the detailed reasons supporting it must then be reviewed with the management team.

When the recommendation is accepted, a plan for "benchmarking" the system selected should be developed to assure the correctness of the decision. The benchmark can be one or two of the company's typical drawings or can take the form of a survey of similar users of the product. Another very good way of "proving" the decision is to have one of the company's designers take the training course provided by the vendor. The cost of the training course should be accepted as money well spent. If the training course confirms the selection, the designer will already be well trained when the system is installed. If the training course should indicate some product negatives, the money spent on training now may save a much greater sum from being spent poorly later. After a final review of the selection which includes reports on the training course, the standard purchasing process of the company should be set in motion and a comprehensive plan of implementation, including site preparation, training management of work flow, considerations for maintenance, and so on, should be developed.

CONCLUSION

A detailed, careful, consistent methodology is important for the selection of a CAD system, just as it is for any computer system or other major capital investment. The reviews, life cycle tasks, and other structure provided by a methodology like SPECTRUM are what makes it work. To the degree that discipline guides following the methodology, it is to that degree that the methodology will work for a particular company. If it works, everyone benefits and the CAD system will be of high quality and the project of selecting and installing it will come in within budget and on schedule.

APPENDIX: CAD HARDWARE AND SOFTWARE CHECKLISTS

TABLE A1 CAD HARDWARE CHECKLIST

		System		
Category	Features	A	B	C
Hardware	Number of component vendors			
	Hardware sources			
	Agents			
General information	General-purpose computer			
	Future upgrade kits			
	Network compatibility			
	Shared peripherals options			
	Single-source service			
	Nationwide service			
	Worldwide service			
	Customized service plans			
	Nearest sales office			
	Nearest service depot			
	Language:			
	Basic			
	Pascal			
	HPL			
	Fortran 77			
	CP/M			
	Other			
	Other software packages available			
	Installation available			

TABLE A1 (Cont.)

		System		
Category	Features	A	B	C
Console	Processor			
	Clock speed			
	Speed arithmetics			
	Standard RAM			
	Maximum RAM			
	Direct memory access (DMA)			
	HPIB interface			
	RS 232 interface			
	Desktop			
	Central processor			
	Network interface			
	RGB monitor interface			
Display	Monochrome			
	Color			
	Type			
	Resolution			
	Size (diagonal)			
	Antiglare filter			
Keyboard	Standard key set			
	Separate number pad			
	User-definable keys			
	Thumb-wheel cursor controls			
	Special-purpose/convenience keys			
Digitizing	Standard tablet			
	Optional tablets			
	Tablet operator			
	Backtrace from plotter			
Mass memory	Flexible disk size			
	Flexible disk capacity			
	Standard hard disk			
	Optional hard disks			
	Magnetic cartridge tapes			
Output	Alphanumeric and graphics thermal printer			
	Alphanumeric and graphics dot matrix printer			
	2-pen 8.5 inch × 11 inch plotter			
	8-pen 11 inch × 17 inch plotter			
	8-pen 24 inch × 48 inch plotter			
	8-pen 36 inch × 48 inch plotter			

TABLE A1   (Cont.)

Category	Features	System A	B	C
	Pens:			
	Transparency			
	Felt			
	Roller-ball			
	Wet-ink			
	Colors			
	Choice of widths			
	Multiple supply source			
	Media:			
	Chart paper			
	Vellum			
	Polyester film (Mylar)			
	Custom sizes			
	Preprinted			
	Special perforation			
	Roll-feed			
	Multiple supply source			
	Maximum plot speed			
	Automatic pen speed/ force defaults			
	Automatic pen selection			
	Automatic pen capping			
	Operator plot override			
	Automatic plot rotation			
	Automatic plot axis align			
	Plotter spooling			
Site preparation	Power per station requirements			
	Maximum power draw (including plotters)			
	Air conditioning/ humidity control requirements			
	Square footage per station requirements			
	Maximum plotter area required			
Costs	Minimum single station, including standard plotter			
	Minimum additional stations, no plotter			
	Minimum special-purpose stations			
	Leasing plan available			
	Lease–purchase plan available			

TABLE A2 CAD SOFTWARE CHECKLIST

		System		
Category	Features	A	B	C
General	Source/manufacturer/distributor			
	Program			
	Language			
	Plain English commands			
	Maximum objects per drawing			
	Power failure data security			
	Operator error checks			
	Source code available			
	Bill-of-materials takeoffs			
	Accessible data base			
	3D graphics			
	Network compatible			
	Nature of data base			
Training	Full, English documentation			
	Built-in tutorials and help functions			
	Training locations			
	Number of operators/ CAD users trained free			
	Number of days of training			
	Cost per extra day/ per extra operator for training			
	In-house training aids available			
Support	Program enhancements available/cost			
	Symbol libraries available			
	Users groups			
	Engineering/architectural/ design consultation			
Automatic drawing aids	Dimension lines and values			
	Crosshatching			
	Fillets			
	Area calculations			
	Perimeter calculations			
	U.S./metric plot scales			
	User-defined plot specifications			
	Zoom			
	Pan			
	Return to view			
	Arrowheads			
Predrawn symbols and details	Predrawn symbol libraries			
	Additions to symbol sets			
	User creation of symbol sets			
	Symbols saved with drawings			
	Maximum number of objects per symbol			
	"Point and place" symbol access			
	Symbol properties override			
	Number of symbols per library			

TABLE A2 (Cont.)

Category	Features	System		
		A	B	C
	Maximum number of libraries/ automatic symbol scaling			
	Switchable symbol levels			
	Symbol alteration/change			
	Automatic library visuals			
	U.S./metric symbol scaling			
Input/display features	Use plotter to digitize old drawings			
	Monochrome or color display			
	Stylus input from standard tablet			
	Thumb-wheel input			
	Keyboard cursor control arrows			
	Absolute coordinate input			
	Relative coordinate input			
	Polar coordinate input			
	Screen scale display			
	User-specified increment snap			
	User-specified grid snap			
	Preimaging of drawing elements			
	Crosshair/cursor sizes			
	User-specified grids			
	Number of drawing levels/layers available			
	Number of screen colors available			
	Object groups independent of levels			
	On/off screen menus and data			
	On/off screen graphics			
	On/off screen grids			
	On/off object center markers			
	On/off crosshatch			
	On/off template lines			
	On/off screen levels			
	User-specified scale/sheet size			
	User-specified active digitizer area			
	Variable object/group handles			
	Feet-inch or decimal input for dimensioning			
	Parallel graphics monitor support			
Output features	Plotter line width/density control			
	Number of pens per plot available			
	User-specified levels to plot			
	Partial drawing plot			
	Archive to hard disk			
	Archive to floppy disk			
	Spooled archive to floppy disk			
	Archive to magnetic cartridge tapes			
	Output alpha/numeric to printer			
	Output graphics to printer			
	Merge drawings			

TABLE A2 (Cont.)

Category	Features	System		
		A	B	C
	English/metric plot scales			
	Plotter sheet composition			
Drawing features	Lines			
	Rectangles			
	Polygons			
	Circles			
	Arcs			
	Ellipses			
	French (Bezier) curves			
	Points			
	Temporary guidelines			
	Symbols/groups			
	Text notes			
Object	Object:			
	Move			
	Copy			
	Copy repeat:			
	1 axis			
	2 axes			
	Circular			
	Axis and rotation			
	Image			
	Image copy			
	Rotate			
	Scale			
	Delete			
	Undelete			
	Explode/dismember			
	Property changes			
	Text:			
	Move			
	Copy			
	Rotate			
	Scale			
	Proportional			
	Unproportional			
	Edit			
	Adjust after rotation			
	Justify			
	Top			
	Bottom			
	Left			
	Right			
	Center			

TABLE A2 (Cont.)

Category	Features	System A	B	C

Group edit	Group:
	  Move
	  Copy
	  Scale
	  Delete
	  Undelete
	  Image
	  Image copy
	  Property changes			

Costs	First workstation initial First workstation renewal Additional workstations initial Additional workstations renewal Software leasing terms available Operator training Updates available/frequency Support			

SUGGESTED READINGS

Auerbach Publishers. *Auerbach on Digital Plotters and Image Digitizers.* Princeton, N.J., 1972.

Barnhill, R. E., and Riesenfeld, R. F. *Computer-Aided Geometric Design.* (Proceedings of Conference at the University of Utah, Salt Lake City, March 1974). New York: Academic Press, 1975.

Beauchemin, R., Krimper, R., Manifor, J., and Sunyogh, J. *CAD-Tutor: A Sequential Teaching Manual for CADAPPLE 2D Drafting System.* Los Angeles: CADventures Unlimited, 1983.

Carpenter, S., Bhupendra, P., and Head, J. The liaison of industry and schools in CADD education. In *Proceedings of the Fourth Annual Conference and Exposition of the National Computer Graphics Association.* Fairfax, Va.: NCGA, 1983.

Chasen, S. H. *Geometric Principles and Procedures for Computer Graphics Applications.* Englewood Cliffs, N.J.: Prentice-Hall, 1978.

Crow, F. C. Three dimensional computer graphics. *Byte Magazine,* March–April 1981.

Design Automation Conference No. 13. *Proceedings, 1976.* San Francisco: IEEE, 1976.

Design Automation Conference No. 14. *Proceedings, 1977.* New Orleans: IEEE, 1977.

Fetter, W. A. *Computer Graphics in Communication.* New York: McGraw-Hill, 1965.

Glowinski, R. *International Symposium on Computing Methods in Applied Science and Engineering.* Berlin: Springer-Verlag, 1976.

Goetsch, D. *The CAD/CAM Workbook.* New York: South-Western, 1983.

Hyman, A. *The Computer in Design.* London: Studio Vista, 1973.

Jarrett, I., and Eavey, L. The revolution in graphic communication fostered by the micro computer. In *Proceedings of the Fourth Annual Conference and Exposition of the National Computer Graphics Association.* Fairfax, Va.: NCGA, 1983.

Jones, G. W., and Smith, D. R. *CAD '76: Proceedings of the Second International Conference on Computers in Engineering and Building Design.* Guilford, England: IPC Science and Technology Press, 1976.

Lazear, T. Low cost CAD. *Proceedings of the First Annual Conference and Exposition of the National Computer Graphics Association.* Fairfax, Va.: NCGA, 1980.

Lazear, T. CAD training. In *Proceedings of the Third Annual Conference and Exposition of the National Computer Graphics Association.* Fairfax, Va.: NCGA, 1982.

Lazear, T. CAD training in industry and education. In *Proceedings of the Fourth Annual Conference and Exposition of the National Computer Graphics Association.* Fairfax, Va.: NCGA, 1983.

Lazear, T. Specifying low cost CAD. *Proceedings of the Fourth Annual Conference and Exposition of the National Computer Graphics Association.* Fairfax, Va.: NCGA, 1983.

Lazear, T. Piping isometrics on personal computers. *Plant Engineering,* December 1983.

McGinnis, R., and Aburdene, M. Integration of interactive graphics and computer aided design into undergraduate engineering curricula. In *Proceedings of the Fourth Annual Conference and Exposition of the National Computer Graphics Association.* Fairfax, Va.: NCGA, 1983.

McGinnis, R., and Hyde, D. The development of computer graphics applications across the curriculum in undergraduate education. In *Proceedings of the Fourth Annual Conference and Exposition of the National Computer Graphics Association.* Fairfax, Va.: NCGA, 1983.

Mitchell, W. J. *Computer-Aided Architectural Design.* New York: Petrocelli-Charter, 1977.

Newman, W. M., and Sproul, R. F. *Principles of Interactive Computer Graphics.* New York: McGraw-Hill, 1973.

Orr, J. *The Complete CAD/CAM Anthology.* Chestnut Hill, Mass.: Management Roundtable, n.d.

Peckham, H. *Computer Graphics.* New York: Scientific Press, n.d.

Prince, D. M. *Interactive Graphics for Computer-Aided Design.* Reading, Mass.: Addison-Wesley, 1971.

Spectrum International. *Spectrum Methodology for Systems Development.* Culver City, Calif., n.d.

Status report of the graphics standards planning committee. *Computer Graphics,* 1977, p. 11.

T&W Systems. *CADAPPLE Designer's Reference Manual.* Huntington Beach, Calif., 1981.

T&W Systems. *3D Designer's Reference Manual.* Huntington Beach, Calif., 1983.

Voisinet, D. D. *Introduction to Computer-Aided Drafting.* New York: McGraw-Hill, 1983.

INDEX